Plato, Diagrammatic Reasoning
and Mental Models

Susanna Saracco

Plato, Diagrammatic Reasoning and Mental Models

Susanna Saracco
Rivoli (TO), Italy

ISBN 978-3-031-27657-6   ISBN 978-3-031-27658-3   (eBook)
https://doi.org/10.1007/978-3-031-27658-3

This Palgrave Macmillan imprint is published by the registered company Springer Nature Switzerland AG.
The registered company address is: Gewerbestrasse 11, 6330 Cham, Switzerland

Contents

List of Figures

1

Introduction

Abstract Plato's words are aimed, not at conveying a static description of how things are, but at creating cognitive stimulations for his readers. I will present the way in which I have chosen to respond to this Platonic *request* for *collaboration*. I am going to analyze the importance of *visualization* for cognitive development. This analysis allows to realize the role of *diagrams* as heuristic devices. Diagrams facilitate problem-solving inferences. *Mental models* will make us understand the mental processes that occur in deductive inference. Via mental models we will begin to turn our rational attention to a superior phase of epistemic development. The intellectual progress in this cognitive phase is facilitated by *ante rem structuralism*, in which numbers are treated as positions in structures.

Keywords Epistemic stimulations • Visual thinking • Diagrammatic reasoning • Mental models • Structuralism

Plato, in the sixth book of the *Republic*, presents his schematization of human intellectual development. Plato explains what are, for him, the stages of cognitive progress of the human being, in connection with the different objects of investigation that the human reason can grasp. Plato schematizes his idea of rational growth using a line segment divided into four subsections: two of them correspond to phases in which our knowledge is still connected to the sensible realm and the other two sectors indicate a kind of knowledge which is pertinent to the intelligible realm. My attention has been seized by the Platonic words, in which the philosopher tells his readers that there is much more to discover on this subject, and this is something that they have to do (*Republic*, VII 534 a). In saying this, Plato calls for a *collaboration* between writer and reader. As Foley points out to us, Plato asks his readers to participate actively with the text. This participation is not meant to be a simple approval or criticism of the words of the philosopher; rather, this call for collaboration is designed to "*force a thoughtful reader to transcend the text*" (Foley 2008, 23. My emphasis). Plato's is a kind of *higher-order pedagogy* that *calls for a collaboration between writer and reader*. Plato has not written a textbook whose content can merely be summarized by the readers. He has created a text to which they are required to respond and the act of responding is as important as the text itself. Plato's readers are invited to become active creators of the philosophical message. This does not mean that Plato's words are incomplete in the sense that they communicate thoughts which have not yet reached a good degree of elaboration. On the contrary, it means that the words written by Plato are so well mastered by their author that they are able to stimulate the reader to overcome them, as Foley was highlighting.

Through the dialogues, Plato is inviting us to reflect on *our* cognitive resources to develop them autonomously. He says this explicitly in the passage of the *Meno* in which he tells us that learning is a process of "*recollection*" (*Meno*, 81 d). It is useful to read these lines together with an excerpt from the *Phaedrus*, in which Plato says to his readers that they have to find the truth *by themselves*, using what they are reading only as a *reminder* of the rational power that they possess (*Phaedrus*, 275a–b). In the *Phaedrus* we read that the written words will not help us to remember but they can only be used as *reminders* because they do not lead to ourselves but they rather depend on signs that "belong to others" (*Phaedrus*, 275 a). In the *Phaedrus* Plato explicitly connects the process of *learning* with remembering something that is *inside* us: what is inside us makes us *remember, recollect,* a wisdom that is merely reminded by the written words.

It seems unlikely that the author of these passages would conceive of his own written words as the final destination of knowledge, but rather as a stimulus to reach that destination, which is internal to us. Thus, the Platonic words are only a *reminder* of the necessity of looking for knowledge where the answers to the dialogical questions come from, *inside* us, in the organ capable of remembering which is, for Plato, the soul and its main component, the reason. Consistently, Plato's dialogues do not end with the thoughts of the author and the words, the *reminders*, that he has selected to convey them, but they are enriched by the multitude of rational memories prompted by the autonomous investigations of Plato's readers.

The courage of recognizing the existence of an intellectual dimension in which what we have learned to consider certain becomes criticizable, losing its stability, is the necessary premise to reconstruct creatively a truth, that is far from the shadows of what merely appears as true, as Mattéi makes us understand, distinguishing *"two sorts of spectacle lovers,"* those who love to watch and those who love to listen (Mattéi 1988, 79. My emphasis). Mattéi highlights that the spectacle created by Plato must not be seen as something constructed to be passively watched and it is not the final destination of the intellectual growth of the reader. *If we confuse a means of rational growth with the final goal of this process,* we are condemned to live in an epistemic realm in which the shadows are for us the reality, in which we merely watch and we do not listen. In this cognitive dimension we will never know the truth. If we recognize that Plato's words compose a succession of *epistemic stimulations* devised to encourage rational evolution, whose meaning requires to be completed by the critical and creative contributions of his readers, we allow the words of Plato to perform the real show they were invented for, the show in which the absolute protagonist is human reason.

My arguments about Plato's idea of cognitive progress entail that Plato's readers play a vital role in determining Plato's message. But this does not imply that this message can be interpreted in just any way by its readers. Even though the individual contributions of Plato's readers cannot be anticipated, this does not mean that there is an unlimited range of possibilities available to them. Their contributions have to take place within certain epistemic conditions established by Plato: they must be grounded on detachment from the tangible, as a necessary condition to reach what for the philosopher is the peak of human rational development, the knowledge of the purely intelligible. Nevertheless, within these broad conditions, Plato's readers can decide to criticize Plato's philosophical system, even radically, and develop alternatives to it.

I have chosen to accept the core of Platonic philosophy and I have decided to engage with his words, using them for an investigation in line with his philosophical system. The way that I have chosen to respond to the Platonic request for *collaboration* has resulted in the elaboration of a new theoretical framework for engaging with Plato's dialogues (Saracco 2017). A part of my reconstruction represents the continuation of the schematization of human cognitive development, traced by Plato in the *Republic*: I add four subsections to the line segment used by Plato to represent cognitive progress (*Republic*, VI 509 d–511). In this way, I indicate the necessity to envisage the intellectual journey subject of the dialogues as *one* stage of the journey of human rational growth. Plato exhorts his readers to work more on the diagram of intellectual progress. Nevertheless, he does not provide specific details about this further investigation on rational development. Thus, a part of the theoretical framework that I have developed, is not based on a direct description elaborated by Plato. This is *not* problematic: the strength of the message that I want to convey does not depend on the specific details of the reconstruction of the Platonic account of human intellectual development. A reader who thinks that the last phase of the cognitive individual growth has to be represented using three subsections of the line segment which symbolizes intellectual development, or a reader who disagrees with the sources that I have used to render the idea of what this rational progress is, is assuming the necessity to contextualize Plato's written words in a broader theoretical framework, represented by an extended line segment. This reader, developing this type of criticisms, is also interacting with the Platonic text, accepting the request of collaboration between reader and writer that I have emphasized as fundamental for the philosopher. This kind of criticisms does not undermine but *reinforces* the basics of my work.

My reconstruction of the highest phase of rational development, that is not directly described by Plato, takes on board a piece of scientific method. In science, when there are testable elements which present variations which are not in line with what was theorized about their properties, it is possible, before rejecting the theories about those elements, to hypothesize that the unpredictable variations are generated by other elements, whose existence was not taken into consideration before. This is the way in which in the nineteenth century the planet Neptune was discovered: the motion of Uranus was considerably different from that predicted through the Newtonian gravitational theory. In order to find a solution to this problem it was hypothesized that there should be a previously undetected planet close to Uranus. The attraction between this hypothetical planet and Uranus had to be considered the cause for the departure of Uranus from its initially predicted orbit. Once this hypothesis was assumed to be true, it was possible to test its content, checking with a telescope for the presence of an undiscovered planet. This led to the first sighting of Neptune, saving Newton's gravitational theory (Chalmers 1976, 78).

When a testable element presents anomalies inconsistent with what it is known about its nature, it is assumed that these variations are caused by another element, not directly testable. In my case, the words of the *Republic* on the necessity to continue the inquiry about human rational development (*Republic*, VII 534 a), are the unpredictable effect, provoked on the dialogues and testable as part of them, caused by a further stage of cognitive progress. This additional phase of rational development, not directly testable because not fully described by Plato in the dialogues, is what I define as theoretical adulthood, successive to theoretical childhood.

The schematization of my reconstruction of Plato's account of cognitive growth, is formed by the line segment, with the four stages of rational progress, traced by Plato's words in the *Republic* (*Republic*, VI 509d–511). These four subsections represent *theoretical childhood*. The length of the line segment of the *Republic* is increased by four subsegments, added to represent the stages of development of *theoretical adulthood*. I am using the term *theoretical* having in mind the relation between *theōreō* and *oraō*, which implies *a process of cognition which starts with the vision, instantiated through* physical or *intellectual eyes*. Thus, theoretical childhood will be that stage of cognition in which the speculations are in their childhood because the *intellectual* eyes are not yet looking in the right direction. With the expressions theoretical childhood and theoretical children, I am *not* referring to *real* children and their cognitive development but I am defining phases of *rational* evolution, one intellectually more advanced than the other, coherent with Plato's indications. The cognitive progress of theoretical children is facilitated by the use of *natural language*. *Different levels of mathematical complexity* justify the attribution to this subject of two cognitive tasks: mathematics promotes the cognitive growth of theoretical children and it has also a central role in the intellectual development of theoretical adults. These levels of mathematical complexity can be understood in association with *two axiomatic approaches*. The mathematics which facilitates the rational development of theoretical children has to be related to the axiomatic approach that I define as top-down: results are obtained deductively starting from premises which are never questioned by the user. A more sophisticated axiomatic approach, that I define as bottom-up, is based on premises which can be questioned in light of the results obtained. This latter axiomatic approach is connected with the intellectual progress of theoretical adults.

This outline of my interpretation of Plato's representation of intellectual development, gives a sense of the *interdisciplinary* nature of my research. This interdisciplinarity characterizes Plato's philosophy, as it is confirmed by the fact that Plato's work is analyzed by researchers whose studies are not focused on ancient philosophy. Marcus Giaquinto's research on the *epistemological* importance of visual thinking in mathematics (Giaquinto 2007) takes into consideration the case of the *Meno*. According to Giaquinto, it is important to recognize the *epistemic role* of visualization: visualization makes us *discover* because it *triggers* "belief-forming dispositions" (Giaquinto 2007, 12). Giaquinto notices that in some cases, as in the case of the *Meno*, the mode of belief-acquisition is fast, thus the resulting belief seems to the subject immediate and obvious (Giaquinto 2007, 47. My emphasis). In the case of the *Meno*, one gets the belief almost immediately, that is, "without any subjectively noticeable period between visualizing and getting the belief. *Immediacy* suggests that to explain why *visualizing* leads to the *belief* we should look to the *visualizer's prior cognitive state*. One hypothesis is that the subject's prior cognitive state included tacitly believing B. *This kind of view was proposed by Plato. On Plato's view the experience of visualizing triggers retrieval of the tacit belief B*" (Giaquinto 2007, 60. My emphasis). In the case of the *Meno*, visualization triggers *immediately* the relevant belief-forming dispositions. This entails that the subject's prior cognitive state already included those dispositions. This is in line with what Plato states about recollection in the *Meno*: we have seen that, according to Plato, the individuals possess wisdom within themselves. For the philosopher the process of learning is memory, *recollection*, of what is *inside* us.

Thanks to the activation of belief-forming dispositions, we acquire concepts, such as square or diagonal, which allow us to discover, as in the case of the slave in the *Meno*, geometrical truths. The use of *diagrams* in the *Meno* has been studied by Giaquinto (1993), as an example of the *epistemological* role of visualization. Giaquinto notes that in a type of process like that which occurs thanks to the dialogical exchange between Socrates and Meno's slave, *the use of diagrams* cannot be a *superfluous adjunct to a proof*, because *no* construction of a *proof of the theorem* followed. Moreover, he pointed to the fact that the *visual experience* that resulted from the *use of diagrams* was, in this case, a means of getting information about things that were *not* before one's eyes (Giaquinto 1993, 95). Thus, the exchange between Socrates and Meno's slave, creates an opportunity for Plato's readers to reflect on the possibility of a *use of diagrams* that does not fall for the erroneous empirical view. It would be wrong to assimilate the *epistemic* role of visualization into an evidence-providing role. The dialogue between Socrates and Meno's slave, gives us the chance to consider the *epistemological* importance of visualization: thinking visually, *discoveries* about geometry can be made.

This awareness of the use of *diagrams* to make us *discover* leads us back to the line segment of the *Republic* and makes us reflect on the *lack of investigation* of the reasons why Plato has chosen *a* schematization to represent intellectual progress and on the reasons why Plato has chosen *that* schematization to represent intellectual progress. My research on *diagrammatic reasoning* helps us to ponder on the first question. A possible answer for the second question is provided by the last part of my research, on theoretical adulthood and structuralism.

Diagrams are helpful to draw inferences useful for *problem solving* (Larkin and Simon 1995). Diagrams are also useful to gain a deeper understanding of the *nature* of the *problem*. This is a fundamental step in the solution process. There are problems that are certainly easier to solve thanks to a diagrammatic representation, rather than listening to an explanation. For example, given the two sentences, "A is shorter than B" and "C is longer than B," anyone can conclude that "A is shorter than C" based on the meaning of the words "shorter" and "longer." However, I can reach the same conclusion much easier and faster by looking at a drawing of three line segments properly labelled and aligned. This advantage of diagrams is not limited to spatial problems; this would be evident if the same drawing of line segments that we have just used to compare lengths, would be used to compare prices of goods (Iwasaki 1995, 657, 658).

One crucial feature of reasoning with diagrams is that it is in general *qualitative*. In the above example of length comparison, even if the exact length of the three line segments were given and the diagram drawn in the correct proportion, what could be discerned just by glancing at the diagram would still be qualitative facts such as ordering relations. Qualitative analysis is of crucial importance, since it allows the understanding of the *global characteristics* of the issue at stake, without being burdened by unimportant details; in this way, can be quickly recognized those places which allow further, more sophisticated, analysis (Iwasaki 1995, 659).

This comparison of the length of line segments, and the consideration of the ordering relations among them, brings back to us the schematization traced by Plato in the *Republic* to represent the different stages of human cognitive development. According to my interpretation, this representation has not been properly interrogated. In fact, I think that the *qualitative* observation that should be made about the line segment of the *Republic*, is *not* about ordering relations. This is *not*, for me, the *global characteristic* that Plato wanted to point out to us. My interpretation of the *nature of the problem*, that can be investigated via the line segment of the *Republic*, is going to be sketched shortly.

In problem solving, figuring out how to represent the information at hand is often the most important part of the solution. The use of *text* and the use of *diagrams* can both lead to inferences which solve the problem taken into consideration. What is crucial in carrying out a reasoning task, is the capability to select the most appropriate form of representation for the reasoning task at hand (Barwise and Etchemendy 1995, 212–213). Efficient reasoning is *heterogeneous*: the search for *any* universal scheme of representation, linguistic or diagrammatic, is a mistake (Barwise and Etchemendy 1995, 212).

A truly heterogeneous inference system, where reasoning uses both language and diagrams, is *Hyperproof*. Hyperproof is used to teach elementary logic courses. Both diagrams and sentences are made available in Hyperproof. Because of the differences in expressive power between diagrams and sentences, neither of the two forms of representation which characterize Hyperproof is made redundant by the other. That is, there are things that can be depicted by diagrams that cannot be said in the language, and vice versa (Barwise and Etchemendy 1995, 218–220). Stenning (2002) helps us to familiarize with Hyperproof: the designers of Hyperproof posed reasoning problems using graphical and sentential information. The givens of the problem are in a diagram window, which is accompanied by sentences in a lower window. Stenning provides an example of the different types of *abstraction* in Hyperproof (Stenning 2002, 56). The reasoning problems posed via Hyperproof improve the *students' general reasoning abilities*. Hyperproof, with its combinations of diagrams and sentences, helps students to see that *logic* is an *abstract* account of representations (Stenning 2002, 62).

We have seen that Plato, in the *Republic* (VI 509 d–511), gives indication to trace a line segment which is the symbol of intellectual development; he also describes the cognitive faculties which work in each rational step, indicating what are the objects which can be grasped at each reasoning phase. Using Hyperproof's language, we can say that this excerpt of Plato's work, provides his readers both with a diagram window, and with sentential information. Plato has chosen to stimulate the rationality of his readers using both diagram and language: he has created an opportunity for the human mind to use different kinds of representation systems, and the different kinds of abstraction, related to them. Plato has posed a *logic* riddle for his readers: they have to investigate on the *nature* of rationality. In doing so, they improve their general rational abilities, learning to learn. This logic stimulation of the readers' general reasoning abilities is the *qualitative* character that we should notice in the line segment traced by Plato in the *Republic*: the *global characteristic* that is pertinent to this schematization is the fact that it is part of a reasoning problem which aims at stimulating via *logic* Plato's readers general reasoning abilities. Also the lines of the *Republic* in which Plato, summarizing his idea of intellectual progress, tells his readers that there is much more to know about the subject than what had been discussed so far with Glaucon (*Republic*, VII 534 a), should be examined in the perspective offered by Hyperproof, as part of the *logic* exercise devised by Plato to facilitate the cognitive development of his readers.

After having seen that Plato has used *a* schematization to make us aware of our rational skills via logic, my research continues with the analysis of *mental models*. This work should be helpful to get a sense of *what kind of inferences* are useful to approach the *logic* problem that Plato poses to favor our rational growth. Plato, describing the stages of human cognitive evolution (*Republic*, VI 509d–511), poses a *logic* problem to stimulate the development of the general reasoning abilities of his readers. A crucial step of this cognitive progress is the moment in which Plato's readers pass from an inferior phase of rational growth, that I have called theoretical childhood, to a superior phase of cognitive development, that I have labeled as theoretical adulthood. This fundamental passage is rendered possible, in my interpretation of Plato, by the Forms. I associate the Forms with scientific models because of their *epistemic function*. Both the Forms and models are not abstract direct representations (ADR): they are *not* abstracted *directly* from the *empirical or the intelligible realm*. They are the *cognitive artifacts*, the mediators, that lead us toward the intelligible. The analogy between Plato's Forms and modeling is constructed thanks to the study of *mental models*, in the notion elaborated by Johnson-Laird (1983, 1988).

Mental models, for Johnson-Laird, make us understand the mental processes that occur in *deductive inference* (Johnson-Laird 1983, 23–24). As Johnson-Laird points out, "*vision* yields mental models" (Johnson-Laird 1988, 231. My emphasis). We have noted, thanks to Giaquinto's research, the important *epistemic role* of *visualization*, which is not a mere source of evidence, but it is what triggers the belief-forming dispositions that render possible new discoveries. The vision involved in the formation of mental models has *not* an evidence-providing role: we construct mental models to ease a process of *deductive inference*. Mental models increase our *logical competence*, improving our capacity for making inferences. The concept of vision related to mental models has an *epistemic* rather than an *empiric* character. We know that, for Plato, the truth can be found only in the realm of the *intelligible*. Thus, mental models are the *media* that, in a Platonic perspective, lead us from an inferior *cognitive* state, that I have labeled as theoretical childhood, to an advanced epistemic phase, that I have called theoretical adulthood. Mental models can be considered *analogues of* the Platonic Forms because of their *epistemic function*: they are both *media* in which *vision* has an *epistemic* role, increasing our cognitive ability in *problem solving*. This emphasis on the kind of *vision considered epistemically* is provided by Plato himself, who chooses the term Idea, also rendered in translation as Form, to point to what is pertinent to the realm of the intelligible. Ideas are intelligible (τὰς δ' αὖ ἰδέας νοεῖσθαι μέν *tas d au ideas noeisthai men Rep*, VI 507b): the word *Idea* comes from the root *-id*, that is found in the verbal form *eidon*, aorist of the verb *oraō*. The meaning of this verb comprises a metaphysical shade: *oraō* means *mental* sight (Liddell et al. 1996, 1245). Thus, Plato chooses to refer to the Forms using a word which is *etymologically related to a concept of vision cognitively tainted*. The vision of the Ideas is a vision that makes you *know*, connecting you with the intelligible realm. Having in mind the relation between *theōreō and oraō*, and the connection between *cognition and vision*, I have chosen, as we have seen, to label the two phases of my reconstruction of cognitive development in Plato, theoretical childhood and theoretical adulthood.

In the last part of this research, I investigate the possible reasons why Plato has chosen *that* diagram, a line segment subdivided into sectors, to stimulate us intellectually. Plato has chosen to represent rational development using *that* diagram to point to us the importance of *mathematics* in the process of rational growth. *Numerals* are associated with a *spatial line*: it is more common to teach *arithmetical operations* in terms of their *spatial analogues* than in terms of direct logical definitions (Sloman 1995, 17). Furthermore, the infinity of the natural number *structure* can be rendered via a mental number line with no right end, one that continues rightward endlessly (Giaquinto 2007, 227). We have distinguished between *two levels of mathematical complexity*: the first level, "the method of geometry and mathematics in general" (Heath 1921, 290), can be associated with an axiomatic approach that we defined as top-down: with this method, results are logically deduced from unquestioned axioms. This level of mathematical complexity is proper to theoretical children. The mathematics utilized by theoretical adults is based on a bottom-up axiomatic approach: at this level of sophistication, the consequences of the problem have to be utilized to consider the truth of the premises.

This higher level of mathematical complexity is represented by Shapiro's *ante rem structuralism*. *Ante rem* structuralism is considered by Shapiro "a variant of traditional Platonism" (Shapiro 2011, 130). According to *structuralism, numbers* should be treated as *positions in structures*. For the structuralist, "mathematics is seen as the investigation…of 'abstract structures', systems of objects fulfilling certain structural relations among themselves and in relation to other systems, without regard to the particular nature of the objects themselves….the 'objects' involved serve only to mark 'positions' in a relational system; and the 'axioms' governing these objects are thought of, *not* as *asserting definite truths*, but as *defining* a type of structure of mathematical interest" (Hellman 2005, 536–537). *Ante rem* structuralism is a kind of structuralism that ignores the *individual properties of the objects*, that are irrelevant, and it considers only an object as a *position* in a structure. Shapiro connects *ante rem* structuralism with Plato's philosophy: for Plato truth is disentangled from the *empirical* realm and can be found in the *purely intelligible*; in the same way, the "*ante rem* structuralist takes a Platonic view of structures: they exist and are available for mathematical description as complex objects in their own right, *whether or not exemplified by any independent collection of objects*" (Wright 2000, 330. My emphasis). For Shapiro, it is irrelevant the *empirical* existence of objects that exemplify the structures that he is taking into consideration; these objects exist *ontologically*, as those *positions* in a structure that can be grasped via an act of *intellection*. *Both for Shapiro and for Plato, the truth is not in the empirical but in the intelligible realm*. The existence of the structures is posited by Shapiro via an axiomatic theory of structures. Shapiro's structures are axiomatically characterized (Sereni 2019, 253); nevertheless, Hellman has just clarified that the axioms, governing the objects that in structuralism are positions in a structure, do *not* assert *definite* truths but they *define* a kind of structure of mathematical interest (Hellman 2005, 537). The axiomatic approach connected to structuralism can be thus related to the axiomatic approach that has been called as bottom-up, based on premises which can be questioned in light of the results obtained.

As we have just seen, *ante rem* structuralism is a theory about what (mathematical) *universals* there are. According to MacBride, there is a crucial epistemological problem that Shapiro has to face: how can a *physical* being located in a physical universe know the *abstract* realm, that includes *ante rem* universals and infinite structures (MacBride 2008)? According to Shapiro, the goal of his research is to demonstrate that mathematical knowledge *just* is knowledge of *ante rem* structures. This has not to be proved from *non*-mathematical premises (Shapiro 2011, 149). Both Shapiro and Plato do not tell us where their universal evidence comes from. But Plato has chosen to provide us with cognitive stimulations which are *entrance points* to this epistemic realm. The cognitive awareness acquired thanks to this rational stimulation gives us the chance to choose to criticize, even radically, Plato's system and every aspect that characterizes it.

References

Texts and Translations

Plato. *Meno*. 1997. Translated by Grube, G.M.A. In *Plato: Complete Works*, ed. J.M. Cooper. Indianapolis: Hackett.

Plato. *Phaedrus*. 1997. Translated by Nehamas, Alexander and Woodruff, Paul. In *Plato: Complete Works*, ed. J.M. Cooper. Indianapolis: Hackett.

Plato. *Republic*. 1997. Translated by Grube, G.M.A. Revised by Reeve, C.D.C. In *Plato: Complete Works*, ed. J. M. Cooper. Indianapolis: Hackett.

Recent Works

Barwise, Jon, and John Etchemendy. 1995. Heterogeneous Logic. In *Diagrammatic Reasoning: Cognitive and Computational Perspectives*, ed. Janice Glasgow, N. Hari Narayanan, and B. Chandrasekaran. Cambridge, MA: MIT Press.

Chalmers, A.F. 1976. *What Is this Thing Called Science?* Indianapolis: Hackett.

Foley, R. 2008. Plato's Undividable Line: Contradiction and Method in Republic VI. *Journal of the History of Philosophy* 46 (1): 1–23.

Giaquinto, Marcus. 1993. Diagrams: Socrates and Meno's Slave. *International Journal of Philosophical Studies* 1 (1): 81–97.

———. 2007. *Visual Thinking in Mathematics: An Epistemological Study.* Oxford: Oxford University Press.

Heath, T. 1921. *A History of Greek Mathematics.* Oxford: The Clarendon Press.

Hellman, Geoffrey. 2005. Structuralism. In *The Oxford Handbook of Philosophy of Mathematics and Logic*, ed. Stewart Shapiro. Oxford: Oxford University Press.

Iwasaki, Yumi. 1995. Introduction to Section IV: Problem Solving with Diagrams. In *Diagrammatic Reasoning: Cognitive and Computational Perspectives*, ed. Janice Glasgow, N. Hari Narayanan, and B. Chandrasekaran. Cambridge, MA: MIT Press.

Johnson-Laird, Philip N. 1983. *Mental Models: Towards a Cognitive Science of Language, Inference, and Consciousness.* Cambridge: Cambridge University Press.

———. 1988. *The Computer and the Mind: An Introduction to Cognitive Science.* Cambridge, MA: Harvard University Press.

Larkin, Jill H., and Herbert A. Simon. 1995. Why a Diagram Is (Sometimes) Worth Ten Thousand Words. In *Diagrammatic Reasoning: Cognitive and Computational Perspectives*, ed. Janice Glasgow, N. Hari Narayanan, and B. Chandrasekaran. Cambridge, MA: MIT Press.

Liddell, Henry G., Robert Scott, Henry Stuart Jones, and Roderick McKenzie. 1996. *A Greek-English Lexicon.* Oxford: Clarendon Press.

MacBride, Fraser. 2008. Can *Ante Rem* Structuralism Solve the Access Problem? *The Philosophical Quarterly* 58 (230): 155–164.

Mattéi, Jean-François. 1988. The Theatre of Myth in Plato. In *Platonic Writings/Platonic Readings*, ed. Charles L. Griswold. New York: Routledge.

Saracco, Susanna. 2017. *Plato and Intellectual Development: A New Theoretical Framework Emphasising the Higher-Order Pedagogy of the Platonic Dialogues.* Cham: Palgrave Macmillan.

Sereni, Andrea. 2019. On the Philosophical Significance of Frege's Constraint. *Philosophia Mathematica* 27 (2): 244–275.

Shapiro, Stewart. 2011. Epistemology of Mathematics: What are the Questions? What Count as Answers? *The Philosophical Quarterly* 61 (242): 130–150.

Sloman, Aaron. 1995. Musings on the Roles of Logical and Nonlogical Representations in Intelligence. In *Diagrammatic Reasoning: Cognitive and Computational Perspectives*, ed. Janice Glasgow, N. Hari Narayanan, and B. Chandrasekaran. Cambridge, MA: MIT Press.

Stenning, Keith. 2002. *Seeing Reason: Image and Language in Learning to Think.* Oxford: Oxford University Press.

Wright, Crispin. 2000. Neo-Fregean Foundations for Real Analysis: Some Reflections on Frege's Constraint. *Notre Dame Journal of Formal Logic* 41 (4): 317–334.

2

The *Collaboration* Between Writer and Reader

Abstract Plato's is a kind of higher-order pedagogy in which the readers are not the passive receptors of a content but they discover themselves as authors of the content. I have chosen to respond to the Platonic intellectual stimulation, proposing a new theoretical framework for engaging with Plato's dialogues: I add four subsections to the line segment used by Plato in the *Republic* to represent cognitive development. I have chosen to call this extension theoretical adulthood, to define a phase of rational evolution more advanced than the previous one, theoretical childhood. The intellectual progress of theoretical children is promoted by natural language; I also distinguish two levels of mathematical complexity: one contributes to the epistemic growth of theoretical children, the other makes theoretical adults evolve intellectually.

Keywords Higher-order pedagogy • Theoretical framework • Natural language • Mathematics

© The Author(s), under exclusive license to Springer Nature Switzerland AG 2023   **17**
S. Saracco, *Plato, Diagrammatic Reasoning and Mental Models*,
https://doi.org/10.1007/978-3-031-27658-3_2

2.1 Plato and the Rational Engagement of His Readers

In my work on Plato and intellectual development (Saracco 2017), I analyzed a crucial passage of the Platonic dialogues: the passage of the sixth book of the *Republic* (*Republic*, VI 509d–511) in which Plato explains what are, for him, the stages of intellectual development of the human being and what are the objects of knowledge pertinent to each phase of cognition. Plato schematizes his idea of intellectual progress using a line segment divided into four subsections: two of them correspond to phases in which our knowledge is still connected to the sensible realm and the other two sectors indicate a kind of knowledge which is pertinent to the intelligible realm.

My attention was captivated by the moment in which Plato, summarizing his idea of cognitive progress, tells his readers that there is much more to know about the subject than what had been discussed so far with Glaucon (*Republic*, VII 534 a):

> But as for the ratios between the things these are set over and the division of either the opinable or the intelligible section into two, let's pass them by, Glaucon, *lest they involve us in arguments many times longer than the ones we have already gone through.* (My emphasis)

Foley (2008, 23), commenting on the previous excerpt from the *Republic*, emphasizes:

> the passage shows that Plato is not willing to set forth his views on the further complexities that have emerged. It is a task that he *intentionally* leaves for his readers, revealing that his final assessment of the role of the divided line is to *force a thoughtful reader to transcend the text*. One significant aspect of the divided line is exactly that Plato refuses to explain its point. (Foley 2008, 23. My emphasis)

Foley's words reveal a crucial insight: Plato's text is a stimulus for a rational investigation which is not meant to *end* in the written words of his dialogues. Plato asks his readers to participate actively with the text. This participation is not meant to be a simple approval or criticism of the words of the philosopher; rather, this call for collaboration is designed to *"force a thoughtful reader to transcend the text"* (Foley 2008, 23. My emphasis). Plato, presenting in the *Republic* his schematization of intellectual development, in connection with the objects of investigation that human reason can grasp, *tells* his readers that there is more to discover on the subject, and this is something that *they* have to do. In saying this, Plato *calls for a collaboration between writer and reader*. Plato has not written a textbook whose content can merely be summarized by the readers. He has created a text to which they are required to respond and the act of responding to the text is as important as the text itself: the two of them together complete Plato's task. Plato does not want to convey a static description of how things are. He has created a text that calls out for completion by the readers' further contributions. This does not mean that Plato's words are incomplete in the sense that they communicate thoughts which have not yet reached a good degree of elaboration. On the contrary, it means that the words written by Plato are so well mastered by their author that they are able to stimulate the reader to overcome them, as Foley was highlighting. Plato's texts are not only composed by words which have the goal of expressing the thinking of their author but they also comprise the thinking of their users.

Through the dialogues, Plato is inviting us to reflect on *our* cognitive resources to develop them autonomously. He says this explicitly in the *Meno*:

> As the whole nature is akin, and the soul has learned everything, nothing prevents a man, after recalling one thing only-a process men call learning-discovering everything else for himself, if he is brave and does not tire of the search, for searching and learning are, as a whole, recollection. (*Meno*, 81 c–d)

It is useful to read these lines together with an excerpt from the *Phaedrus*, where Socrates is reporting a dialogue about the art of writing which takes place between Thamus and Theuth:

O most expert Theuth, one man can give birth to the elements of an art, but only another can judge how they can benefit or harm those who will use them. And now, since you are the father of writing, your affection for it has made you describe its effects as the opposite of what they really are. In fact, it will introduce forgetfulness into the soul of those who learn it: they will not practice using their memory because they will put their trust in writing, which is external and depends on signs that belong to others, instead of trying to remember from the inside, completely on their own. You have not discovered a potion for remembering, but for reminding; you provide your students with the appearance of wisdom, not with its reality. Your invention will enable them to hear many things without being properly taught, and they will imagine that they have come to know much while for the most part they know nothing. And they will be difficult to get along with, since they will merely appear to be wise instead of really being so. (*Phaedrus*, 275 a–b)

Let us consider this passage in connection with the passage of the *Meno* cited above: in the *Meno* Plato tells us that learning is a process of "recollection" (*Meno*, 81 d) and in the *Phaedrus* we read that the written words will not help us to remember but they can only be used as *reminders* because they do not lead to ourselves but they rather depend on signs that "belong to others" (*Phaedrus*, 275 a). In the *Phaedrus* Plato explicitly connects the process of learning with remembering something that is *inside* us: what is inside us makes us remember, recollect, a wisdom that is merely reminded by the written words.

It seems unlikely that the author of these passages would conceive of his own written words as the final destination of knowledge, but rather as a stimulus to reach that destination, which is internal to us. Thus, the Platonic words are only a *reminder* of the necessity of looking for knowledge where the answers to the dialogical questions come from, *inside* us, in the organ capable of remembering which is, for Plato, the soul and its main component, the reason. Consistently, Plato's dialogues do not end with the thoughts of the author and the words, the *reminders*, that he has selected to convey them, but they are enriched by the multitude of rational memories prompted by the autonomous investigations of Plato's readers.

The courage of recognizing the existence of an intellectual dimension in which what we have learned to consider certain becomes criticizable, losing its stability, is the necessary premise to reconstruct creatively a truth, that is far from the shadows of what merely appears as true, as Mattéi makes us understand in the following quote, distinguishing "*two sorts of spectacle lovers*" (Mattéi 1988, 79. My emphasis):

> The first are the crowd and the sophists who unreservedly dedicate them- selves to the sensible beauty of colors, forms and voices. As Socrates puts it to Glaucon: '*those who love to watch*' (φιλοθεάμονες) and 'those who love to listen' (φιλήκοοι; *Rep.*, 475 d2) remain the *prisoners of appearances* even if they show an unconscious desire for a higher kind of knowledge. In front of them, '*those who love to know*'-*the philosophers*-are in search of the luminous theater of *truth beyond the shadow play*. Like the pure souls released from their bodies and contemplating the vast plain of Truth, and like the initiate in Eros' mysteries contemplating the boundless ocean of the Beautiful, 'the genuine philosophers are those who are in love with the spectacle of the truth'. (*Rep.*, 475 e)

Here Mattéi highlights that the spectacle created by Plato must not be seen as something constructed to be passively watched and it is not the final destination of the intellectual growth of the reader. If we confuse a means of rational growth with the final goal of this process, we are con- demned to live in an epistemic realm in which the shadows are for us the reality. In this cognitive dimension we will never know the truth. If we recognize that Plato's words compose a succession of epistemic stimula- tions devised to encourage rational evolution, whose meaning requires to be completed by the critical and creative contributions of his readers, we allow the words of Plato to perform the real show they were invented for, the show in which the absolute protagonist is human reason.

2.2   Rational Engagement and Higher-Order Pedagogy

The dialogical character of Plato's work is opposite to the will of indoctri- nating or just instructing the readers. Plato *chose* to write *dialogues* and this choice is not only a formal but also a philosophical choice: Plato wants to stimulate an active participation of his readers which goes beyond the accidental criticism of the written words, that can take place whenever a text is read.

In fact, as we have just seen, when Plato in the *Republic*, has presented his idea of what intellectual development is, he states explicitly that there is more to discover on the subject, but he does not tell his readers *how* they should do it. The modes of collaboration between writer and reader advocated by Plato are not predetermined by the philosopher. Plato's readers can choose to criticize radically his philosophical system or they can choose to accept its basics. Plato interacts dialogically with his readers, asking them explicitly to transcend the text (Foley 2008, 23. Also cf. *Phaedrus*, 275 a–b) to complete it with their contributions. This Platonic request is at the base of the *higher-order pedagogy* that permeates the dialogues, where the role of the readers is not flattened to that of students who can merely absorb the content proposed by their teacher. Plato's readers are invited to become active creators of the philosophical message. This invitation has not to be considered as a consequence of a lack in Plato's argumentative ability. On the contrary, as we have just seen, the philosopher is able to stimulate his readers with explicit requests.[1]

For Plato, education has crucial importance. In fact, the philosopher is well aware of the fact that the human rational nature can diverge from its positive capabilities, when its direction is determined by messages that appeal simply to appetite. This intuition is itself remarkable for its modernity. But what renders the Platonic rational pedagogy extraordinary is its character: Plato explicitly says to his readers that they have to find the truth by themselves, using what they are reading only as reminder of the rational power that they possess (*Phaedrus*, 275 a–b). Plato's is a kind of higher-order pedagogy in which the readers are not the passive receptors of a content but they discover themselves as authors of the content.

The dialogue between Plato and his readers takes place via the written words of his texts, that allow the continuation of the cognitive exchange between the philosopher's rational heritage and his reader's intellect. The dialogical interaction with the readers, and the consequent free development of their thinking abilities, does not mean that the Platonic philosophy can be developed in any way. The intellectual stimulation of Plato's words consists in the exhortation to contribute in an original and creative way to the development of what Plato thinks that knowledge is. Plato tells his readers clearly what his idea of knowledge is: the highest point of intellectual development is reached when we are able to abandon the empirical completely to reach the purely intelligible. Only when our rationality is disentangled from the distracting stimuli which come from the tangible realm, we are able to grasp the purely intelligible truth. Nonetheless, the individual contributions of Plato's readers can mould the concept of Platonic knowledge into the shape their intellect suggests. Furthermore, it remains possible at any point for Plato's readers to use their rational capabilities, sharpened through the texts written by the philosopher, to criticize his conception of knowledge, abandoning in this way Plato's philosophical system. My work does not go in this direction. I have chosen to respond to the Platonic intellectual stimulation, proposing a new theoretical framework for engaging with Plato's dialogues.

2.3   Plato's Higher-Order Pedagogy: My Response

I am going to present the basics of the new theoretical lens that I have elaborated as response to the Platonic request to collaborate with his text. I have chosen to accept the core of Platonic philosophy and I have decided to engage with his words, using them for an investigation in line with his philosophical system. At the center of my engagement with Plato's words there is the account of human intellectual development presented in the *Republic* (*Republic*, VI 509 d–511), schematized using a line segment divided into four subsections:

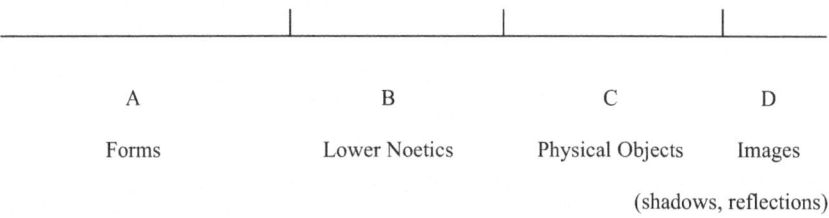

This is the rendition, chosen by Foley, of Plato's discussion of the progress of the cognitive capacities of the individual. Each object indicated in the line segment above can be apprehended thanks to a rational faculty correspondent to it (Foley 2008, 1). The subsection A corresponds to Understanding, *noēsis*. At this stage of intellectual development the individual is able to apprehend the Forms. The subsection B is Thought, *dianoia*. In this phase of rational evolution the person begins his investigation of the mathematical objects, intellectually inferior to the Forms. The subsection C is Belief, *pistis*, which gives the person the chance to understand the physical objects. The subsection D is Imagination, *eikasia*, which is used to know the images. As Foley explains, he has preferred to "follow one general tendency in the literature of labeling the section representing the Forms with the letter 'A' and treating it as the longest subsegment because Forms are first in order of importance"(Foley 2008, footnote 1, p. 1).

The different length of the subsections of the line segment is traditionally used to represent the different cognitive importance of the objects which correspond to them and of the intellectual faculties necessary to understand these objects. Longer subsections represent objects more difficult to grasp and more advanced cognitive faculties, necessary to investigate these objects. Foley comments on the lines of the *Republic* quoted in Sect. 2.1, in which Plato exhorts his readers to investigate further the subject of human cognitive progress, stating that even if it seems that the Platonic indications to divide the line segment entail the existence of two middle subsections of *equal* length, when we analyze further this schematization we see that "the two middle subsegments are *unequal* because they represent mental states of unequal clarity, and possibly also objects with unequal degrees of reality"(Foley 2008, 1).

I disagree with Foley because I think that the words of Plato cited above have not to be interpreted only within the cognitive space of the four sectors of the line segment that we have examined. On the contrary, these sectors are the starting point of an intellectual progress which is not described in the dialogues but is originated by them. Plato's words, in my interpretation, are an exhortation to keep in mind that the content of the dialogues is just one chapter of the Platonic book of knowledge.[2] This must guide our interaction with the Platonic text, in case we decide to cooperate with it, as I have done, accepting to stay within the conceptual boundaries given by the Platonic conception of knowledge, which culminates with the apprehension of the purely intelligible. In my reconstruction of what the Platonic account of human intellectual progress could be, I am aware of the role of his written words, in respect to the larger cognitive project that the philosopher indicates. But I am also aware that this broader theoretical framework, even though it has to respond to the Platonic idea of truth, which has to be totally separated from the empirical, leaves us the necessary intellectual space to shape this truth with our contributions.

This positive characteristic of Platonic philosophy leads to the fact that my reconstruction of the stages of human intellectual development,[3] respects and is guided by the Platonic principles about knowledge and truth but it is disputable because it cannot respond to a precise Platonic description. Nevertheless, I need to make an assumption in order to progress with my research on Plato's ideas about human rational growth. I take on board a piece of scientific method to elaborate my theory about what could be the stages of cognitive progress, which should be added to those described in the *Republic*. In science, when there are testable elements which present variations which are not in line with what was theorized about their properties, it is possible, before rejecting the theories about those elements, to hypothesize that the unpredictable variations are generated by other elements, whose existence was not taken into consideration before.

This is the way in which in the nineteenth century the planet Neptune was discovered: the motion of Uranus was considerably different from that predicted through the Newtonian gravitational theory. In order to find a solution to this problem it was hypothesized that there should be a previously undetected planet close to Uranus. The attraction between this hypothetical planet and Uranus had to be considered the cause for the departure of Uranus from its initially predicted orbit. Once this hypothesis was assumed to be true, it was possible to test its content, checking with a telescope for the presence of an undiscovered planet. This led to the first sighting of Neptune, saving Newton's gravitational theory (Chalmers 1976, 78).

In our case, the Platonic excerpt which we have taken into consideration via Foley's comment, is the unpredictable effect which confirms our theory about the existence of stages of cognitive development, which add subsections to the line segment used by Plato to represent human intellectual progress. These subsections are indicated with A′, B′, C′, D′ in the schematization below and they are our Neptune, which has not been noticed before.

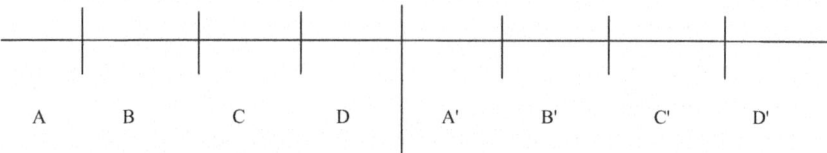

As we have seen, Foley has chosen to represent with A the Forms, pointing at the significance of this object and of the cognitive capacity correspondent to its understanding, through the use of a subsection of the line segment of intellectual progress larger than the others. In my line segment, the subsection A represents the images and the cognitive capacity necessary to grasp them. When we are able to understand D, the Forms, we reach a superior level of intellectual development. Starting from this epistemic moment, we are able to begin the investigation of the purely intelligible, which is for Plato the highest rational achievement. In the dialogues, there is no indication of how this investigation can take place. I have hypothesized that there can be stages of rational progress also in the cognitive development of the individuals who are already able to investigate the purely intelligible. For this reason, I have also hypothesized that the analysis of the purely intelligible has to begin with an empirical aid, as it happens in the first stages of rational development described by Plato. These stages are represented by the subsections A and B of my line segment, that are, as A′ and B′, still related to the empirical. With this notation, I suggest the correspondence between the stages of cognitive development, A–D, necessary to reach the epistemic point in which we are able to start the investigation of the purely intelligible and the stages of cognitive advancement, A′–D′, of the individuals who are already able to research the purely intelligible.

In order to stress that the description of human intellectual evolution given in the *Republic* is only the first part of the cognitive progress of the individual, I have chosen to call the four sectors of the line segment traced in the *Republic*, theoretical childhood (A–D); the extension of this line segment is theoretical adulthood (A'–D'). I am using the term *theoretical* having in mind the relation between *theōreō* and *oraō*, which implies *a process of cognition which starts with the vision, instantiated through* physical or *intellectual eyes*. Thus, theoretical childhood will be that stage of cognition in which the speculations are in their childhood because the intellectual eyes are not yet looking in the right direction. With the expressions theoretical childhood and theoretical children, I am *not* referring to *real* children and their cognitive development but I am defining phases of rational evolution, one intellectually more advanced than the other, coherent with Plato's indications.

Plato states explicitly what are the objects analyzed during the rational progression from A to D. The purely intelligible is the most complex object that the human reason can examine. Thus, it is plausible that its knowledge takes place in stages and that the beginning of the investigation of the purely intelligible is still informed by the tangible, as means to reach the purely intelligible. We do not know whether A', B', C', D' correspond to different objects which reveal different aspects of the purely intelligible or whether different cognitive layers of the purely intelligible are the objects of investigation in A', B', C', D'. But my addition of subsections in the line segment of cognitive progress described by Plato has not the purpose of providing the final answer about the Platonic account of human intellectual development.[4] My representation of this account wants to emphasize that the individual rational growth, as envisaged by Plato, does not end in the description of the *Republic* (*Republic*, VI 509 d–510) but it continues with stages of rational development complementary to those traced by the Platonic words. This extension, grounded in the lines of the *Republic* commented by Foley (*Republic*, VII 534 a), is my way of responding to the request for collaboration with the text which is, as we have seen, a fundamental aspect of the Platonic dialogues.

In my representation, all the sectors of the line segment have equal length.[5] This does not mean that I think that there is no theoretical difference among the objects and mental stages which correspond to the parts of the line segment. In fact, the text of the *Republic* provides fuel for discussion of equal or unequal length of the subsections of the line segment. Joining this discussion would serve no purpose in my interest on this representation of the Platonic account of intellectual development.

This interest is focused on the *equal epistemic* significance that *each* subsection has for the individual rational development. Maintaining the focus on the *function* of each epistemic stage of the line segment is crucial to grasp the significance of this representation for the understanding of the nature and potentiality of human rationality according to Plato.

We have seen so far the phases of development of theoretical childhood and adulthood. Now I want to present the basics[6] of my reconstruction of what could be for Plato the different means, or techniques as I call them, which favor rational progress in each of the phases which are part of theoretical childhood and theoretical adulthood. The technique that Plato has chosen to make theoretical children evolve cognitively is the use of natural language. The beginning of the analysis of this technique is given by the quotation from the *Phaedrus* that we have already taken into consideration in Sect. 2.1. In that excerpt, Socrates reports a dialogue between Thamus and Theuth about the art of writing. For the present purposes, our attention has to be focused on the distinction, made in that excerpt, between knowledge which stems from *external reminders* and *knowledge which emerges exclusively from the reasoning capabilities of the individual*. In Sect. 2.1, quoting the *Meno*, we have spoken about the ability of the *reason* to remember, to *recollect*, originating knowledge by itself. But when we have not yet developed this skill we need the words, external reminders of our cognitive potentialities.

We need only to be *reminded* about our intellectual capacities because even during a phase, theoretical childhood, in which we have not yet reached a high degree of intellectual sophistication, we already possess the skills to attain this goal. This is stressed by Plato in the following lines:

> Education isn't what some people declare it to be, namely, putting knowledge into souls that lack it, like putting sight into blind eyes...*the power to learn is present in everyone's soul*...education...it isn't the craft of putting *sight into the soul*. *Education* takes for granted that sight is there but that it isn't turned the right way...and *it tries to redirect it appropriately*. (*Republic*, VII 518 c–d. My emphasis)

These words are part of Book VII of the *Republic*, where the allegory of the cave shows the necessity that the eyes of those who have always lived in the obscurity of the appearance of knowledge adjust gradually to the sight of its bright reality. This excerpt points to the *graduality* of the process of human intellectual development, as it is confirmed from the context in which these lines appear. The reasoning ability is a skill proper of the human beings and it belongs to every one of them. Nonetheless, to make sure that the cognitive eyes look at the truth, it is necessary that they are appropriately stimulated. This will avoid the danger emphasized by Mattéi in the lines quoted in Sect. 2.1: people stop at the spectacle created by Plato's *words* without investigating its *function*.

We have taken into consideration words to point at their usefulness for the rational growth of theoretical children. Nevertheless, natural language is not the appropriate technique for the rational stimulation of theoretical adults. As we have seen, the object of investigation of theoretical adults is the purely intelligible. To understand what could be an adequate technique to promote the development of this higher-level thinking, I am going to start from Foley's emphasis on the importance attributed by Plato to mathematics. As we have seen in Foley's discussion of his rendition of Plato's line segment which represents objects and the cognitive faculties necessary to understand them, mathematical objects are the first point of entrance in the realm of the intelligible. This is what Foley explains, emphasizing

> the tremendous importance that mathematics has in Plato's account of philosophical development. The study of mathematics serves as a bridge between physical objects and the Forms. Learning to think mathematically is presented as a necessary condition for thinking philosophically because mathematics is what leads us from concern for physical objects to understanding of eternal objects. Once this transition to eternal objects has been made, it is easier to study the Forms. (Foley 2008, 12)

We have stressed the significance of Foley's thought about the Platonic text as stimulation for a research which has not to end with those written words. Now he points at the need of considering the crucial role that mathematics plays in Plato's philosophy, as the bridge between an inferior level of rational development, which can know only via the physical realm, and a superior intellectual refinement, which is able to grasp the non-sensible, the Forms.

I agree with Foley's statements about the significant role that mathematics plays to reach the highest intellectual goal according to Plato, the knowledge of the purely intelligible. Nevertheless, Heath stresses a difference between mathematical and dialectical method in Plato which can make us think that mathematics is imperfect in comparison with dialectic and it cannot be the technique which promotes a higher-order development of human rationality:

> Plato distinguishes two processes: both begin from hypotheses. *The one method cannot get above these hypotheses* but, treating them as if they were the first principles, builds upon them and, with the aid of diagrams or images, arrives at conclusions: *this is the method of geometry and mathematics in general.* The other method treats the hypotheses as being really hypotheses and nothing more, but uses them as stepping-stones for mounting higher and higher until the principle of all things is reached, a principle about which there is nothing hypothetical; when this is reached, it is possible to *descend again*, by steps each connected with the preceding step, to the conclusion, a process which has no need of any sensible images but deals in ideals only and ends in them; this method, which rises above and puts an end to hypotheses, and reaches the first principle in this way is the *dialectical method.* (Heath 1921, 290. My emphasis)

These lines should not be considered as the base for an exclusion of mathematics from the realm of theoretical adulthood. This would be an incorrect inference which can be avoided if we take into consideration the different levels of mathematical complexity.

The first level of mathematical complexity can be associated with an axiomatic approach which can be defined as top-down axiomatic approach. This is "the method of geometry and mathematics in general" (Heath 1921, 290): it helps us to prove that results are correct (Greenberg 1974, 8) using the axioms, which are never questioned by the user,[7] and the logical consequences we derive from them. With this method results are logically deduced from unquestioned axioms, which are the foundations which ground the mathematical structure.[8] Greenberg explains to us what an axiom is, emphasizing that

If I wish to persuade you by *pure reasoning* to believe some statement S1, I could show you how this statement follows logically from some other statement S2 that you may already accept. However, if you don't believe S2, I would have to show you how S2 follows logically from some other statement S3. I might have to repeat this procedure several times until I reach some statement that you already accept, one I do not need to justify. That statement plays the role of an *axiom* (or *postulate*). If I cannot reach a statement that you will accept as the basis of my argument, I will be caught in an "infinite regress," giving one demonstration after another without end. (Greenberg 1974, 9)

Greenberg's words point to the fact that the *axioms* are grasped through an *act of intellection*. This reminds us of the role of mathematics in the redirection of our cognitive sight toward the intelligible, which Foley was emphasizing.

I have pointed at the existence of two levels of mathematical complexity. We have just seen briefly the utility of the geometrical axioms to move from the tangible to the intelligible. This focus on the intelligible is for Plato fundamental to evolve intellectually *till to the point in which we become theoretical adults. The mathematics utilized by theoretical adults*, already emerges from Heath's words about the dialectical method. When mathematics is applied to the understanding of complex problems, it is not anymore based upon axioms, which do not require any reconsideration. On the contrary, at this level of sophistication, the consequences of the problem have to be utilized to reconsider the truth of the premises (Russell 1973, 273–274). In this case, we have not a rational movement which merely goes from an element to its mathematical consideration via a mathematical principle which will not require any reevaluation. This is the way in which the axiomatic approach which we defined as top-down works and its relative simplicity allows its utilization by theoretical children, favoring their cognitive progress toward theoretical adulthood. But, as Foley has highlighted, for Plato the highest point of intellectual evolution is reached when the purely intelligible is the only subject of investigation. At that speculative level, theoretical adults have to try to solve problems whose complexity demands to go back from what has been considered a correct result, a correct consequence of their thinking, to its premise. This axiomatic approach can be called bottom-up since the progress of theoretical adults in the understanding of the consequences of their line of reasoning will illuminate the comprehension of the related premises.

I have indicated that the written words are useful reminders for individuals whose intellectual skills have not yet been totally developed. When Plato's readers reach the cognitive complexity of theoretical adults they have no necessity of the mediation of a written text to progress intellectually. Indeed, this text would be very difficult to compose because it should describe the myriads of intellectual routes which can be chosen by a mind whose capacity of selection is not restrained by cognitive mistakes. This kind of description would be not only very challenging to write but also useless since the only people who could grasp its content would be those who have already reached a level of intellectual maturity which renders the written reminders pointless. This level of development of the human intellectual capacities is not the object of a *direct* Platonic description. Thus, my reconstruction of theoretical adulthood is, in a sense, solidly grounded in Plato's text because it is a reconstruction of a phase of human rational development based, as we have seen, on the effects that this cognitive phase, theoretical adulthood, provokes on another phase, theoretical childhood, directly described by Plato. Nevertheless, the ground of theoretical adulthood is *meant* to be shaken by the contributions of minds which have no fear to leave the place of tradition to develop innovative researches. Consequently, I am ready to admit not only that my idea of theoretical adulthood can be criticizable but also that if it was not criticizable, it would not be that territory of novelty, correspondent to the Platonic choice of leaving this cognitive zone to the rational talent of his readers.

Notes

1. Plato's intellectual stimulations are *not* limited to the *explicit* requests of collaboration between writer and reader that the philosopher introduces in his dialogues. Plato is also able to elaborate intellectual stimulations whose meaning is unveiled gradually by the readers who progress rationally. I define both the explicit and the non-explicit cognitive stimuli

devised by Plato in the dialogues as *epistemic games*. The nature and the features of the *epistemic games* are analyzed in my book, Saracco, S. 2017. *Plato and Intellectual Development: A New Theoretical Framework Emphasising the Higher-Order Pedagogy of the Platonic Dialogues*. Cham, Switzerland: Palgrave Macmillan. See in particular the second chapter, *The Structure of Rational Engagement in the Reading of Plato*, pp. 13–52.

2. Stating this I do *not* want to associate my theory with the point of view of those scholars who claim that Platonic *basic* teachings are not part of his written dialogues because they belong to his unwritten doctrines (See the Tübingen school, in particular Krämer, Hans J. 1990. Edited and translated by Catan, John R. *Plato and the Foundations of Metaphysics: A Work on the Theory of the Principles and Unwritten Doctrines of Plato with a Collection of the Fundamental Documents*. Albany: State University of New York Press and Szlezák, Thomas. 1999. *Reading Plato*. Translated by Zanker, Graham. London: Routledge). On the contrary, I do think that the *fundamental* Platonic teachings *are* in the written dialogues. The existence in this work of indications of the presence of a stage of rational evolution, complementary to the intellectual development rendered possible by the Platonic written texts, does not mean that there are fundamental concepts of Plato's philosophy that are not part of his written words. My idea is that the basics of Plato's thought *are* in the dialogues but the dialogues should not be considered as the final stage of cognitive evolution but as the means to reach a further stage of rational development, whose detailed description is not provided by Plato.

3. On this subject see Saracco, S. 2017. *Plato and Intellectual Development: A New Theoretical Framework Emphasising the Higher-Order Pedagogy of the Platonic Dialogues*. Cham, Switzerland: Palgrave Macmillan. See in particular the third chapter, *Theoretical Childhood and Theoretical Adulthood*, pp. 53–83 and Saracco, S. 2016. "Theoretical Childhood and Adulthood: Plato's Account of Human Intellectual Development." *Philosophia: Philosophical Quarterly of Israel*, 44 (3).

4. This is not problematic: the strength of the message that I want to convey does not depend on the specific details of the reconstruction of the Platonic account of human development. A reader who thinks that the last phase of the cognitive individual growth, that I call theoretical adulthood, has to be represented using three subsections of the line segment which symbolizes intellectual development, instead of the four subsections that I have chosen to represent this phase of cognitive development,

is assuming the necessity to contextualize Plato's written words in a broader theoretical framework, represented by an extended line segment. This reader, developing this type of criticisms, is also interacting with the Platonic text, accepting the request of collaboration between writer and reader that I have emphasized as fundamental for the philosopher. This kind of criticisms does not undermine but reinforces the basics of my work.

5. The equal length of the subsections of my line segment does not aim at suggesting that the ancient Greek text should be revised so that the modified words would create the chance to compose unproblematically the Platonic schematization of the stages of intellectual progress using four equal subsegments ("The Revisionist Interpretation" (Foley 2008, 8–9)). I also do not want to commit myself to the idea that "the two middle segments were not meant to be compared" (Foley 2008, 9–12). This is the way in which the length of the sectors of the line segment of the *Republic* is treated in the so-called demarcation interpretation. Its name derives from the fact that its exponents think that exists a "clear *demarcation* between the intended and unintended points of comparison, and such a demarcation will show that the equality of the middle subsegments can be dismissed because it falls into the latter category"(Foley 2008, 10). I am not interested here in debating whether the equality of the two middle subsegments is unintended ("The Gaffe Interpretation" (Foley 2008, 12–15)), or intended ("The Dissolution Interpretation" (Foley 2008, 15–18)). I want simply to stress the more general point that all the four subsections described in the *Republic* (*Republic*, VI 509 d-511) are important for our cognitive growth but the significance of the process of human intellectual evolution cannot be fully grasped if its reconstruction is limited to these sectors.

6. More on this subject in the third chapter of my book (Saracco 2017), *Theoretical Childhood and Theoretical Adulthood*.

7. Obviously, this does not imply that axioms can never be the subject of criticism.

8. An example of how the axiomatic method works, in connection with its application to solve the first problem of Euclid's *Elements*, can be found in the third chapter of my book (Saracco 2017), *Theoretical Childhood and Theoretical Adulthood*, pp. 70–73.

References

Texts and Translations

Plato. *Meno*. 1997. Translated by Grube, G.M.A. In *Plato: Complete Works*, ed. J.M. Cooper. Indianapolis: Hackett.

Plato. *Phaedrus*. 1997. Translated by Nehamas, Alexander and Woodruff, Paul. In *Plato: Complete Works*, ed. J.M. Cooper. Indianapolis: Hackett.

Plato. *Republic*. 1997. Translated by Grube, G.M.A. Revised by Reeve, C.D.C. In *Plato: Complete Works*, ed. J.M. Cooper. Indianapolis: Hackett.

Recent Works

Chalmers, A.F. 1976. *What is this Thing Called Science?* Indianapolis: Hackett.

Foley, R. 2008. Plato's Undividable Line: Contradiction and Method in Republic VI. *Journal of the History of Philosophy* 46 (1): 1–23.

Greenberg, M.J. 1974. *Euclidean and Non-Euclidean Geometries: Development and History*. San Francisco: W. H Freeman and Co.

Heath, T. 1921. *A History of Greek Mathematics*. Oxford: The Clarendon Press.

Krämer, Hans J. 1990. *Plato and the Foundations of Metaphysics: A Work on the Theory of the Principles and Unwritten Doctrines of Plato with a Collection of the Fundamental Documents*. Edited and translated by John R. Catan. Albany: State University of New York Press.

Mattéi, Jean-François. 1988. The Theatre of Myth in Plato. In *Platonic Writings/Platonic Readings*, ed. Charles L. Griswold. New York: Routledge.

Russell, Bertrand. 1973. The Regressive Method of Discovering the Premises of Mathematics. In *Essays in Analysis* by Russell, Bertrand, ed. Douglas Lackey. London: George Allen & Unwin Ltd.

Saracco, Susanna. 2016. Theoretical Childhood and Adulthood: Plato's Account of Human Intellectual Development. *Philosophia: Philosophical Quarterly of Israel* 44 (3): 845–863.

———. 2017. *Plato and Intellectual Development: A New Theoretical Framework Emphasising the Higher-Order Pedagogy of the Platonic Dialogues*. Cham: Palgrave Macmillan.

Szlezák, Thomas. 1999. *Reading Plato*. Translated by Zanker, Graham. London: Routledge.

3

Visual Thinking

Abstract The use of *diagrams* in the *Meno* has been studied by Giaquinto, as an example of the *epistemological* role of visualization: visualization makes us *discover* because it *triggers* belief-forming dispositions. Thanks to the activation of these belief-forming dispositions we acquire concepts, such as square or diagonal, which allow us to discover geometrical truths. It would be wrong to assimilate the *epistemic* role of visualization into an evidence-providing role: in the dialogical exchange between Socrates and Meno's slave, *the use of diagrams* cannot be a *superfluous adjunct to a proof,* because *no* construction of a *proof of the geometry theorem* followed. Moreover, the *visual experience* that resulted from the *use of diagrams* was a means of getting information about things that were *not* before one's eyes.

Keywords Visual discovery • Belief-forming dispositions

3.1   Visual Thinking in the *Meno*

We have seen that the criticisms of Plato's words are not mere accidents: their occurrence is *provoked* by the dialogical interaction to make them become part of the philosophical message itself. As we said, this epistemic stimulation is not meant to make us accept Plato's idea of truth. We, as readers of the Platonic dialogues, are rationally stimulated by Plato to discover a rational sophistication of which we were not aware. We are guided by someone who knows more than we do, but we are guided by him through a dialogical exchange. This method makes us discover the rational resources which give us the chance to critically evaluate the thoughts of the person who is intellectually guiding us, acquiring at the same time the capability of completing his own system and the independence from its content. Through the dialogues, Plato is inviting us to reflect on *our* cognitive resources to develop them autonomously.

An example of the importance of the investigative freedom of the rational creature is found in the dialogue *Meno*, where Meno's slave will discover that he possesses the intellectual ability to find an answer to a geometrical problem thanks to the dialogical interaction with Socrates. The slave is not pressured to accept the point of view of an earlier theorist or Socrates' beliefs; indeed, Socrates never expresses his point of view but he questions his interlocutor to develop in him the awareness of his intellectual abilities. The cognitive growth of Meno's slave takes place in the fictional stage of the *Meno*: the slave's answers are decided by Plato as part of his fictional creation but this creation points at the importance of the independent rational activity of the subject of a dialogical interaction. Even when the contribution of Meno's slave is limited to an affirmative or negative answer his replies reveal his own rational activity, stimulated by the words of his interlocutor but developed independently from them (see in particular *Meno*, 81 c–e). In fact, the solution of a geometrical problem by someone who has never studied geometry requires a reasoning which, even if it is not fully recorded in the dialogue, is present in the correctness of the slave's answer. Thus, the slave's answers are not perfunctory because they are signaling a process of active reflection, required to reply correctly to the questions presented. In the *Meno* the slave is not questioned to learn Socrates' truth, he is questioned to discover that there is truth in himself.

The slave in the *Meno*, through Socrates' questioning, acquires conscience of his rational abilities but what kind of thinking is involved in the reasoning of the slave who gradually realizes to possess the cognitive capacity to know a geometrical truth? An answer to this question comes from Marcus Giaquinto's research. He has worked on the epistemological importance of visual thinking in mathematics.[1] According to Giaquinto "the oldest and best known discussion of visual *discovery* is to be found in Plato's *Meno* (82b–86b)" (Giaquinto 2008, 32. My emphasis). Giaquinto explains that it is usually considered impossible to *discover* a geometrical theorem thanks to visualization. This happens because, when visualizing and seeing are compared, it is usually felt that visualizing is no better than seeing (Giaquinto 2007, 67). This is due to a misleading comparison: in fact, "while the *experience* of visualizing is similar to the experience of seeing, the *epistemic role* of visualizing can be utterly different from the primary, evidence-providing role of seeing… So the fundamental mistake here is to assume that the epistemic role of visual experience, whether of sight or imagination, must be to provide *evidence*" (Giaquinto 2007, 67. My emphasis).

To understand better the *epistemic* role of visualization according to Giaquinto, it is necessary to come back to the *Meno*. There (*Meno*, 81 e–86 c), Plato famously presented a visual way of *discovering* a simple fact of geometry: if a diagonal of one square is a side of another square, this other square has twice the area of the first (Giaquinto 2007, 12), as illustrated in the diagram below (Fig. 3.1):

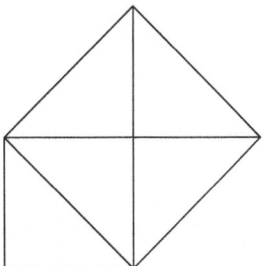

Fig. 3.1 Illustration of a simple fact of geometry: if a diagonal of one square is a side of another square, this other square has twice the area of the first

Giaquinto emphasizes the necessity that every geometrical discovery has a starting point. Thus, the initial challenge of this Platonic visual discovery is this: "how can we acquire *basic* geometrical knowledge?" (Giaquinto 2007, 12). According to Giaquinto

> In having geometrical concepts for shapes, we have certain general *belief-forming dispositions*. These dispositions can be triggered by experiences of seeing or visual imagining, and when that happens we acquire geometrical beliefs. The beliefs acquired in this way constitute knowledge…provided that the belief-forming dispositions are reliable. (Giaquinto 2007, 12. My emphasis)

In this excerpt Giaquinto explains that a visual *discovery* involves the activation of dispositions, that he defines as "belief-forming dispositions" (Giaquinto 2007, 12) that come with possession of certain geometrical concepts (e.g., square, diagonal). What *triggers the activation* of these dispositions is conscious *visual experience*. A belief acquired in this way is *non-empirical*, "because *the role of experience is not to provide evidence*. At the same time, some *visual experience is essential for activating the relevant belief-forming disposition*" (Giaquinto 2007, 47. My emphasis). Giaquinto notices that in some cases, as in the case of the *Meno*, the mode of belief-acquisition is fast, thus the resulting belief seems to the subject immediate and obvious (Giaquinto 2008, 33). In very many cases we are unaware of the cause and occasion of the acquisition of a belief. In fact,

> having a belief is not a manifest state like a pain state-some of our beliefs we are unaware of having-and the transition from lacking a certain belief to having it may also occur without awareness…One may not get a firm belief all at once; to acquire a firm belief by activation of a belief-forming disposition, activations on several occasions may be needed. *But the point is unchanged: there is no anomaly in the fact that we are usually unaware of those occasions.* (Giaquinto 2007, 39. My emphasis)

In the case of the *Meno*, one gets the belief almost immediately, that is, "without any subjectively noticeable period between visualizing and getting the belief. *Immediacy* suggests that to explain why *visualizing* leads to the *belief* we should look to the *visualizer's prior cognitive state*. One hypothesis is that the subject's prior cognitive state included tacitly believing B. *This kind of view was proposed by Plato. On Plato's view the experience of visualizing triggers retrieval of the tacit belief B*" (Giaquinto 2007, 60. My emphasis).

Giaquinto's research has helped us to see in the *Meno* an example of visual *discovery*. Visual thinking is based upon visual activation of belief-forming dispositions. Thanks to the activation of these belief-forming dispositions we acquire concepts, such as that of square or diagonal, which allow us to discover, as in the case of the slave in the *Meno*, geometrical truths. In the case of the *Meno* visualization triggers *immediately* the relevant belief-forming dispositions. This entails that the subject's prior cognitive state already included those dispositions. This is in line with what Plato states about recollection in the *Meno*: in Chap. 2, Sect. 2.1 we have seen that, according to Plato, the individuals possess wisdom within themselves. For the philosopher the process of learning is memory, *recollection*, of what is *inside* us.

For Giaquinto visualization has *epistemic* importance since its role is *not* that of providing evidence; rather, visualization, activates the relevant belief-forming dispositions which render possible a visual discovery. This epistemological role of visual thinking has contributed to make us realize *how* the Platonic dialogues can stimulate cognitively the readers; in fact, the words of the *Meno* promote in Plato's readers an *epistemic* progress via *visual thinking*.

3.2   Meno's Slave and Diagrams

We have just seen that Socrates' questions to Meno's slave have the goal to make the slave realize *his* own intellectual ability. Moreover, we have realized how the dialogical exchange between Socrates and Meno's slave creates an occasion for the rational growth of Plato's readers. In Chap. 2, Sects. 2.1 and 2.2, I propose an interpretation of Plato's words as cognitive stimulations of the readers which contribute to their intellectual development, requiring a collaboration with the text which is not a mere acceptance of its words. The following considerations, written by Giaquinto, are in line with this hermeneutic horizon:

by following the text supplemented by diagrams, one can *discover for oneself* the geometric theorem as it might have been discovered by the slave if he had complied with Socrates' request to give as answers only what he genuinely believed (83d2) rather than what he guessed Socrates believed; or, if one already knows the theorem, one can see how it *could* be discovered that way by someone not already in the know…So one can approach the text as if the purpose of the exchange between Socrates and the slave is to acquaint or re-acquaint readers with the relevant phenomena of *discovery* through their own experience in following the text. (Giaquinto 1993, 82. My emphasis)

by following the exchange and looking at appropriate diagrams *you can yourself go through* a process of a kind that would lead to *discovery* if the theorem were new to you. You can imagine that *the slave's route to discovery is a process of just the same type. And you can ask of just that type of process, a process you actually went through in following the text to reach the theorem, how diagrams are used in it.* (Giaquinto 1993, 86. My emphasis)

In these lines, Giaquinto notices that the readers of the *Meno*, following the dialogical exchange between Socrates and Meno's slave, realize that *they for themselves* are able to *discover* a geometrical truth as the slave did, thanks to the intellectual stimulation of Socrates' questioning. Plato's readers are also cognitively stimulated to reflect on a process of discovery *of the same type*, considering *how diagrams are used in it.* The *Meno* gives his readers an opportunity to grow intellectually, thinking about the possibility to discover a geometric theorem using diagrams. Working on *how diagrams are used* to *discover* a geometric truth, Plato's readers collaborate *actively* with his text. This active engagement with the text, is a fundamental characteristic of Plato's higher-order pedagogy, as we have seen in Chap. 2, Sects. 2.1 and 2.2.

Reflecting on the dialogue between Socrates and Meno's slave, Plato's readers can consider how a geometric *discovery* can be made, and they can reason about the role that diagrams have in this discovery. Giaquinto interrogates himself as to whether every use of a diagram in mathematics falls into one of these two kinds: diagrams used as sources of empirical evidence and diagrams used as helpful but superfluous accompaniments to a proof (Giaquinto 1993, 81). He notices that an alternative use of diagrams is possible, as it is exemplified by the type of process which leads to the understanding of a general theorem of geometry via Socratic questioning. In this type of process,

> the use of diagrams in this process is *clearly not a superfluous adjunct to a proof* (a valid sequence of sentences), since *no proof of the theorem was followed or constructed*. On the other hand, the *use of diagrams was not empirical*: *the visual experience* that resulted from the *use of diagrams* was *not* used as a *source of observational evidence* for this or that proposition. In this case *vision was a means of getting information about things that were not before one's eyes*. (Giaquinto 1993, 95. My emphasis)

Giaquinto notes that in a type of process like that which occurs thanks to the dialogical exchange between Socrates and Meno's slave, *the use of diagrams* cannot be a *superfluous adjunct to a proof*, because *no* construction of a *proof of the theorem* followed. Moreover, he pointed to the fact that the *visual experience* that resulted from the *use of diagrams* was, in this case, a means of getting information about things that were *not* before one's eyes. Thus,

> the exchange between Socrates and the slave in Plato's *Meno*, approached in the right way, reveals another role of diagrams: in this role they enable us to *think visually* about mathematical subject matter in a way which can be assimilated neither to gathering visual evidence nor to picturing a situation independently described in a proof. (Giaquinto 1993, 81. My emphasis)

The exchange between Socrates and Meno's slave, creates an opportunity for Plato's readers to reflect on the possibility of a *use of diagrams* that does not fall for the erroneous empirical view. As we have seen in Sect. 3.1, it would be wrong to assimilate the *epistemic* role of visualization into an evidence-providing role. The dialogue between Socrates and Meno's slave, gives us the chance to consider the *epistemological* importance of visualization: thinking visually, *discoveries* about geometry can be made.

We have just realized that Plato uses diagrams to make us *discover*, growing intellectually. In Chap. 2, Sect. 2.3 we have seen a diagram, traced by Plato in the *Republic*, whose importance is crucial, since it illustrates the different phases of the cognitive development of the human being. I have already stated that, in my opinion, the line segment that schematizes intellectual progress, has not been questioned in the right way. The considerations about visual thinking and the use of diagrams in the *Meno* should have led us back to the line segment of the *Republic* and should have made us reflect on the *lack of investigation* of the reasons why Plato has chosen *a* schematization to represent intellectual progress and on the reasons why Plato has chosen *that* schematization to represent intellectual progress. I am going to work on this investigation in the next part of this research.

Notes

1. Giaquinto, Marcus. 2007. *Visual Thinking in Mathematics: An Epistemological Study*. Oxford: Oxford University Press. For geometrical knowledge see in particular Chaps. 2, 3 and 4. See also Giaquinto, Marcus. 2008. "Visualizing in Mathematics." In *The Philosophy of Mathematical Practice*, edited by Mancosu, Paolo. New York: Oxford University Press.

References

Text and Translation

Plato. *Meno*. 1997. Translated by Grube, G.M.A. In *Plato: Complete Works*, ed. J.M. Cooper. Indianapolis: Hackett.

Recent Works

Giaquinto, Marcus. 1993. Diagrams: Socrates and Meno's Slave. *International Journal of Philosophical Studies* 1 (1): 81–97.
———. 2007. *Visual Thinking in Mathematics: An Epistemological Study*. Oxford: Oxford University Press.
———. 2008. Visualizing in Mathematics. In *The Philosophy of Mathematical Practice*, ed. Paolo Mancosu. New York: Oxford University Press.

4

Diagrammatic Reasoning

Abstract *Diagrams* are helpful to draw inferences useful for *problem solving*. Efficient reasoning is *heterogeneous*. A truly heterogeneous inference system is Hyperproof: the givens of the problem are in a *diagram* window, which is accompanied by *sentences* in a lower window. We have seen that Plato, in the *Republic*, represents intellectual development via a *line segment*; he also *describes* the cognitive faculties pertinent to each rational step, and the objects which can be grasped at each reasoning phase. Using Hyperproof's language, we can say that this excerpt of Plato's work provides his readers both with a diagram window, and with sentential information. Plato has posed a *logic* riddle for his readers: they have to investigate on the *nature* of rationality.

Keywords Problem solving • Heterogeneous reasoning • Hyperproof

4.1 Diagram as Heuristic Device

In discussing the issue of the role of diagrams in reasoning, it is useful to distinguish between *external* diagrammatic representations and *internal* diagrams. The *external* diagrammatic representations are constructed by

S. Saracco, *Plato, Diagrammatic Reasoning and Mental Models*,
https://doi.org/10.1007/978-3-031-27658-3_4

the agent in a medium in the external world, such as paper. The *internal* diagrams comprise the internal representations that are not stored on paper but are held in human memory (Chandrasekaran et al. 1995, XVII).

Larkin and Simon (1995) work on the *use of diagrams* in *problem solving*. In particular, they analyze the reasons why a diagram can be superior to a verbal description for solving problems. They point to the fact that diagrams can group together all information that is used together, in this way they reduce search because related elements are usually close together. This avoids large amounts of search for the elements needed to make a *problem-solving inference*. Thus, diagrams facilitate perceptual *inferences* and recognition of problem-solving methods that may be applicable. Moreover, diagrams minimize labeling: information about an element is near it. Also, diagrams allow quick checks that the analysis is proceeding correctly.

Even though Larkin and Simon's work is focused on the use of *external* representations in problem solving, they also comment briefly on the role of mental images in problem solving. Mental imagery is an *internal* use of diagrams, which are not traced on paper, but are kept in the human mind. According to Larkin and Simon

> *mental images* play a *role* in *problem solving* quite analogous to the role played by *external* diagrams...By this we mean that mental images, while containing substantially less detail than can be stored in external diagrams, have similar properties of localization of information and can be accessed by the same inference operators as the external diagrams. This implies also that the *creation of a mental imagine* (for instance, from a verbal description) employs *inference processes* like those that make information explicit in the course of *drawing a diagram*.
>
> Thus, *when we draw* a rectangle and its two diagonals, the existence of the point of intersection of the diagonals is inferred automatically-the point is created on the paper, accessible to perception. In exactly the same way, *when we imagine* a rectangle with its two diagonals, we imagine ("*see*") the point of intersection in memory. (Larkin and Simon 1995, 105–106. My emphasis)

The distinction between external and internal diagrammatic representations has only been sketched, it will be outlined further in the part of this research dedicated to mental models. For the moment, we should focus on the possibility created by the *use of diagrams* to *draw inferences* useful for *problem solving*. External and internal diagrammatic representations share the same *role*: they promote cognitive development, giving us the chance to draw inferences which make us solve problems. This is exemplified with reference to the processes of mental arithmetic: when you multiply two numbers, you

> construct a model in *your mind's eye* of the *visual* stages that accompany pencil and paper solutions: the *external* symbolic conventions are made available in the *mind's eye* as an *internal* tool for thought. (Bibby 1992, 160. My emphasis)

We still do not know what it is made available to the mind of Plato's readers via the line segment that, in the *Republic*, schematizes intellectual progress. Via this diagram we know that there are phases of rational growth and that there are different cognitive objects that we can grasp in each phase. But what is the problem that Plato wants that we solve? What kind of inferences can be helpful to approach this riddle? The next steps of this research should add new pieces to this cognitive puzzle.

4.1.1 Diagrams and Problem Solving

Diagrams are often used to solve problems. Drawing an appropriate diagram, in which the relations given in a problem are correctly represented, is often helpful to gain deeper understanding of the *nature of the problem*. This is a fundamental step in the solution process. There are problems that are certainly easier to solve thanks to a diagrammatic representation, rather than listening to an explanation. For example, given the two sentences, "A is shorter than B" and "C is longer than B," anyone can conclude that "A is shorter than C" based on the meaning of the words "shorter" and "longer." However, I can reach the same conclusion much easier and faster by looking at a drawing of three line segments properly labeled and aligned. This advantage of diagrams is not limited to spatial problems; this would be evident if the same drawing of line segments that we have just used to compare lengths, would be used to compare prices of goods (Iwasaki 1995, 657, 658).

One crucial feature of reasoning with diagrams is that it is in general *qualitative*. In the above example of length comparison, even if the exact length of the three line segments were given and the diagram drawn in the correct proportion, what could be discerned just by glancing at the diagram would still be qualitative facts such as ordering relations. Qualitative analysis is of crucial importance, since it allows the understanding of the *global characteristics* of the issue at stake, without being burdened by unimportant details; in this way, can be quickly recognized those places which allow further, more sophisticated, analysis (Iwasaki 1995, 659).

This comparison of the length of line segments and the consideration of the ordering relations among them, brings back to us the schematization traced by Plato in the *Republic* to represent the different stages of human cognitive development. In Chap. 2, Sect. 2.3 I have stressed that, according to my interpretation, this representation has not been properly interrogated. In fact, I think that the *qualitative* observation that should be made about the line segment of the *Republic*, is *not* about ordering relations. This is *not*, for me, the *global characteristic* that Plato wanted to point out to us. My interpretation of the *nature of the problem*, that can be investigated via the line segment of the *Republic*, will gradually emerge with the development of this research.

As we have just seen, diagrams are not helpful only to work on spatial problems. There are many cases in which a non-visual problem is aided by visual analogs: mappings from temporal phenomena to spatial phenomena are very common; in fact, we often represent time as a line and reason with lengths of lines when we want to compare durations (Chandrasekaran et al. 1995, XXIV). In human thinking there is often a rich interplay between different forms of representation: it is more common to teach arithmetical operations in terms of their spatial analogues than in terms of direct logical definitions. It is often found useful to think of numbers as forming a spatially ordered series, along which something can move in either direction. Thus, it is sometimes helpful, e.g., for a child learning to understand numbers, to think of subtraction as moving backwards a fixed number of steps (Sloman 1995, 17). People naturally map abstract concepts such as time and numbers to space. Graphical displays capitalize on these natural mappings (Hegarty and Stull 2012, 623). Diagrams are an important aid to work on non-visual problems. We are going to investigate whether Plato used the line segment delineated in the *Republic* to solve a visual or non-visual problem.

4.2 Heterogeneous Reasoning

We have just noticed how often human reasoning relies on different forms of representation to solve problems. Non-visual problems such as arithmetical operations are aided by visual analogs as in the case in which we think about subtraction as moving backward a fixed number of steps in a spatially ordered series formed by numbers. Efficient reasoning is *heterogeneous*: the search for *any* universal scheme of representation, linguistic or diagrammatic, is a mistake. In problem solving, figuring out how to represent the information at hand is often the most important part of the solution. The use of text and the use of diagrams can both lead to inferences which solve the problem taken into consideration. What it is crucial in carrying out a reasoning task, is the capability to select the most appropriate form of representation for the reasoning task at hand (Barwise and Etchemendy 1995, 212–213).

A search for a universal representation system in effective reasoning is misguided. A truly heterogeneous inference system, where reasoning uses both language and diagrams, is Hyperproof. Hyperproof is used to teach elementary logic courses. Both diagrams and sentences are made available in Hyperproof. Because of the differences in expressive power between diagrams and sentences, neither of the two forms of representation which characterize Hyperproof is made redundant by the other. That is, there are things that can be depicted by diagrams that cannot be said in the language, and vice versa (Barwise and Etchemendy 1995, 218–220).

4.2.1 Hyperproof

Stenning considers how the "major modalities of representation-*diagram and language*- are chiefly differentiated as *tools of reasoning* by the ways that they express *abstraction*" (Stenning 2002, 52. My emphasis). *Abstraction* is a vital requirement in a *representation system* but abstraction works differently in diagrams and sentences. Sentences have an abstract syntax which breaks up their representation into a sequence, each member insulated from the others. Diagrams have no such syntax, thus an interpreted spatial relation between their symbols is interpreted between *all* their symbols (Stenning 2002, 36). This distinction between abstraction in diagrams and abstraction in sentences does not mean that these systems are isolated from each other. On the contrary, Stenning agrees with Barwise and Etchemendy (1995): reasoning goes on in systems that result from combinations of modalities-heterogeneous systems (Stenning 2002, 56).

Hyperproof is a computer environment, designed by Barwise and Etchemendy, for teaching first-order logic in a novel way that uses both graphical and linguistic representations-heterogeneous reasoning (Stenning 2002, 54–55). The designers of Hyperproof posed reasoning problems using graphical and sentential information. The givens of the problems are in a diagram window, which is accompanied by sentences in a lower window. Stenning shows a diagram window, which gives an example of the different types of abstraction in Hyperproof (Stenning 2002, 56) (Fig. 4.1):

Two blocks appear off to the side of the board. These indicate *not* that there are blocks *off* the board, but rather that there are blocks on the board *in addition* to the ones which are shown there. Off-board icons may or may not indicate size and shape but they *abstract over position* (Stenning 2002, 55). These abstraction tricks are essential for *posing reasoning problems*. These reasoning problems improve the *students' general reasoning abilities.* Hyperproof, with its combinations of diagrams and sentences, helps students to see that *logic* is an *abstract* account of

Fig. 4.1 "Graphical abstraction in Hyperproof. This diagram contains symbols of varying degrees and types of abstraction...The unlabelled cylinder with the question mark badge on the board lacks size, shape and label attributes, but still has a position; its twin off the chequerboard lacks even a position. The neighbouring medium sized dodecahedron labelled d lacks only a position" (Stenning 2002, 56)

representations. Moreover, introducing diagrams enables the solution of much more complex problems than can be addressed in elementary logic course using sentential reasoning (Stenning 2002, 56, 59, 62).

There is a frequently encountered argument among logic teachers *against* the introduction of the kind of semantics Hyperproof provides. Hyperproof is a *partially interpreted* language. Its predicates have *pre-assigned* meanings. Nevertheless, partially interpreted formalisms are the *didactic ladder* which can be thrown away when the student has ascended it, but without any aid to the climb, the vast majority never grasps anything (Stenning 2002, 62. My emphasis).

In Chap. 2, Sect. 2.3. we have seen that Plato, in the *Republic* (*Republic*, VI 509d–511), gives indication to trace a line segment which is the symbol of intellectual development; he also describes the cognitive faculties which work in each rational step, indicating what are objects which can be grasped at each reasoning phase. Using Hyperproof's language, we can say that this excerpt of Plato's work, provides his readers both with a diagram window, and with sentential information. Plato has chosen to stimulate the rationality of his readers using both diagram and language: he has created an opportunity for the human mind to use different kinds of representation systems, and the different kinds of abstraction, related to them. Plato has posed a *logic* riddle for his readers: they have to investigate on the nature of rationality. In doing so, they improve their general rational abilities, learning to learn.

The final lines of Chap. 3, Sect. 3.2., pointed to the lack of investigation of the reasons why Plato has chosen *a* schematization to represent intellectual progress: Hyperproof has given us the chance to reflect on the possibility that Plato has chosen *a* schematization to represent human intellectual faculties, as part of a *logic* problem, (cf. Sect. 4.1., where it has been asked what could be the problem that Plato wanted to solve using diagrams), which aims at using both the representation systems, diagram and sentences. This logic stimulation of the readers' general reasoning abilities is the *qualitative* character (Sect. 4.1.1) that we should notice in the line segment traced by Plato in the *Republic*: the *global characteristic* that is pertinent to this schematization is the fact that it is part of a reasoning problem which aims at stimulating via *logic* Plato's readers general reasoning abilities.

In the example above of a diagram window in Hyperproof, we have seen that two blocks appear off to the side of the checkerboard. They create the occasion to work on an abstraction *over* position, reflecting on the fact that there are blocks on the board *in addition* to the ones that are shown there. We have noticed in Chap. 2, Sect. 2.1 that Plato, summarizing his idea of intellectual progress, tells his readers that there is much more to know about the subject than what had been discussed so far with Glaucon (*Republic*, VII 534 a):

> But as for the ratios between the things these are set over and the division of either the opinable or the intelligible section into two, let's pass them by, Glaucon, *lest they involve us in arguments many times longer than the ones we have already gone through.* (My emphasis)

It has been pointed to Foley (2008, 23)'s comment on these lines of the *Republic*, that highlights that

> the passage shows that Plato is not willing to set forth his views on the further complexities that have emerged. It is a task that he *intentionally* leaves for his readers, revealing that his final assessment of the role of the divided line is to *force a thoughtful reader to transcend the text.* One significant aspect of the divided line is exactly that Plato refuses to explain its point. (Foley 2008, 23. My emphasis)

We have stressed that Foley's words reveal a crucial insight: Plato *calls for a collaboration between writer and reader.* Plato asks his readers to participate actively with the text. This participation is not meant to be a simple approval or criticism of the words of the philosopher; rather, this call for collaboration is designed to "*force a thoughtful reader to transcend the text*" (Foley 2008, 23. My emphasis). The Platonic words, that follow the summary of his idea of cognitive growth, can be compared to blocks that appear off to the side of the checkerboard in the diagram window in Hyperproof: they offer to Plato's readers a *logic* exercise, calling them for an active collaboration with the text. This collaboration with Plato's text, gives the readers the chance to sharpen their reasoning abilities realizing, at the same time, that they possess these rational skills.

As we have seen, the blocks that appear off to the side of the checker-board in Hyperproof are *not* indications that there are blocks *off* the board; rather, they indicate that there are blocks on the board *in addition* to the ones which are shown there. I do not want to associate my interpretation of Plato with the point of view of those scholars who claim that Platonic basic teachings are *not* part of his written dialogues because they belong to his unwritten doctrines. Rather, I have read the lines in *Republic*, VII 534 a as indicators that there are blocks on the board *in addition* to the ones which are shown in *Republic*, VI 509d–511: as we have seen (Chap. 2, Sect. 2.3), I disagree with Foley because I think that Plato's words in *Republic*, VII 534 a, have not to be interpreted only within the cognitive space of the four sectors of the line segment that have traditionally been examined but they have to be read in respect to the larger cognitive project that the philosopher indicates.

Obviously, the cognitive stimulation of his readers via logic has been devised by Plato. Thus, in a sense, it is a *partially interpreted* language, as it happens with the pre-assigned meanings which characterize the semantics of Hyperproof. But, as Stenning has pointed to us, this *didactic ladder* can be thrown away when the student has ascended it, but without any aid to this climb, the vast majority of the students never grasp anything (Stenning 2002, 62. My emphasis). In our case, when the Platonic cognitive stimulation has made us progress intellectually, we can decide to use the rational capabilities that we have sharpened through the Platonic texts, to criticize, even radically, his conception of knowledge (Chap. 2, Sect. 2.2).

We have seen that Plato has used *a* schematization to make us aware of our rational skills via logic; in the last part of this research, we are going to investigate the possible reasons why Plato has chosen *that* diagram (Chap. 3, Sect. 3.2), a line segment subdivided into sectors, to stimulate us intellectually. In the next section of this research, we are going to work on mental models. This should help us to get a sense of what kind of inferences (Sect. 4.1) are useful to approach the logic problem that Plato poses to favor our rational growth.

References

Text and Translation

Plato. *Republic*. 1997. Translated by Grube, G.M.A. Revised by Reeve, C.D.C. In *Plato: Complete Works*, ed. J.M. Cooper. Indianapolis: Hackett.

Recent Works

Barwise, Jon, and John Etchemendy. 1995. Heterogeneous Logic. In *Diagrammatic Reasoning: Cognitive and Computational Perspectives*, ed. Janice Glasgow, N. Hari Narayanan, and B. Chandrasekaran. Cambridge, MA: MIT Press.

Bibby, Peter A. 1992. "Mental Models, Instructions and Internalization." In *Models in the Mind: Theory, Perspective and Application*, edited by Rogers, Yvonne, Andrew Rutherford, and Peter. A. Bibby. London: Academic Press.

Chandrasekaran, B., Janice Glasgow, and N. Hari Narayanan. 1995. Introduction. In *Diagrammatic Reasoning: Cognitive and Computational Perspectives*, ed. Janice Glasgow, N. Hari Narayanan, and B. Chandrasekaran. Cambridge, MA: MIT Press.

Foley, R. 2008. Plato's Undividable Line: Contradiction and Method in Republic VI. *Journal of the History of Philosophy* 46 (1): 1–23.

Hegarty, Mary, and Andrew T. Stull. 2012. Visuospatial Thinking. In *The Oxford Handbook of Thinking and Reasoning*, ed. Keith J. Holyoak and Robert G. Morrison. Oxford: Oxford University Press.

Iwasaki, Yumi. 1995. Introduction to Section IV: Problem Solving with Diagrams. In *Diagrammatic Reasoning: Cognitive and Computational Perspectives*, ed. Janice Glasgow, N. Hari Narayanan, and B. Chandrasekaran. Cambridge, MA: MIT Press.

Larkin, Jill H., and Herbert A. Simon. 1995. Why a Diagram Is (Sometimes) Worth Ten Thousand Words. In *Diagrammatic Reasoning: Cognitive and Computational Perspectives*, ed. Janice Glasgow, N. Hari Narayanan, and B. Chandrasekaran. Cambridge, MA: MIT Press.

Sloman, Aaron. 1995. Musings on the Roles of Logical and Nonlogical Representations in Intelligence. In *Diagrammatic Reasoning: Cognitive and Computational Perspectives*, ed. Janice Glasgow, N. Hari Narayanan, and B. Chandrasekaran. Cambridge, MA: MIT Press.

Stenning, Keith. 2002. *Seeing Reason: Image and Language in Learning to Think.* Oxford: Oxford University Press.

5

Mental Models

Abstract The analysis of *mental models* helps to get a sense of *what kind of inferences* are useful to approach the *logic* problem that Plato poses when he points to the stages of cognitive evolution. The fundamental passage from an inferior phase of rational growth to a superior stage of cognitive development is rendered possible, in my interpretation of Plato, by the Forms. I associate the Forms with scientific models because of their *epistemic function*. Both the Forms and models are not abstract direct representations: they are *not* abstracted *directly* from the *empirical or the intelligible realm*. They are the *cognitive artifacts*, the mediators, that lead us toward the intelligible. The analogy between Plato's Forms and modeling, is constructed thanks to the study of *mental models*.

Keywords Cognitive artifacts • Abstract direct representation • Deductive inference

© The Author(s), under exclusive license to Springer Nature Switzerland AG 2023   **57**
S. Saracco, *Plato, Diagrammatic Reasoning and Mental Models*,
https://doi.org/10.1007/978-3-031-27658-3_5

5.1   Plato's Forms as Mediators

In Chap. 2, Sect. 2.3 I have outlined the new theoretical framework, that I have elaborated to render Plato's idea of human intellectual growth. This cognitive progression has been illustrated using the line segment traced by Plato in the *Republic* (VI 509d–511). I have added to this schematization four complementary subsections and I have specified that this extension represents theoretical adulthood. Theoretical adulthood is a phase of rational growth subsequent to theoretical childhood. In my reconstruction, natural language and mathematics are the two means, or techniques as I have defined them in Sect. 2.3, used by Plato to stimulate the rational evolution of the individual. This cognitive growth does not correspond to the understanding of a definitive message transmitted by Plato through the words of his dialogues. As we have seen, the dialogues are used by the philosopher to require the active intellectual participation of his readers. Their thinking is crucial to complete Plato's message. This does not mean that Plato's philosophy is incomplete; rather, this signifies that its distinctive traits go together with its non-definitive nature: Plato's thought requires the contribution of the readers with no imposition of a particular line of reasoning as the one that the readers have to follow. The readers are those who choose whether they want to develop their rational investigations respecting the boundaries of Plato's philosophy, which is centered on the abandonment of the empirical realm to reach the purely intelligible. My contribution to the Platonic philosophy goes in this direction: I have focused my research on the reconstruction of Plato's thinking. This is not the only option available to Plato's readers: Plato's readers can use the advancement of their intellectual skills, promoted by Plato's text, to criticize Plato's ideas, developing a different philosophy.

Plato's higher-order pedagogy (Chap. 2, Sect. 2.2) does not teach a definitive message: Plato does not write a textbook with the purpose of making us merely absorb its content; on the contrary, he uses his words to make us discover our rational capabilities. This intellectual progress can result in the departure from Plato's philosophy. In this philosophy, the purely intelligible is considered the peak of human cognition. As we have seen (Sect. 2.3), the purely intelligible is the object of investigation of theoretical adults. *The Forms allow Plato's readers to pass from an advanced stage of development of theoretical childhood to theoretical adulthood.* To get a sense of how this is possible, I am going to use an analogy between the Forms and scientific modeling. At the base of this association, there is the fact that crucial characteristics of the Forms match with fundamental traits of scientific models. Nevertheless, the features that ground the analogy between Forms and scientific modeling, should not be our main focus, if we want to grasp the significance of this analogy. The association between Forms and models is important because it creates the possibility of rendering with precision the *epistemic function* of the Forms, as that *cognitive tool* crucial to investigate indirectly the purely intelligible. The investigation of the purely intelligible is for Plato the highest goal that the human mind can have. Scientific models are modern tools for the investigation of the empirical. Associating Forms and models I do not want to change their respective ultimate objectives: the different final destinations of Forms and models are not an obstacle for their association, which is based on the fact that both Forms and models are *cognitive artifacts* devised to ease the *indirect* research on complex phenomena.

There is no widespread consensus on how to bind the category "cognitive artifacts." Nevertheless, the prototypical cases seem clear: cognitive artifacts are artificial devices made by humans for the purpose of aiding, enhancing, or improving cognition. Examples of cognitive artifacts include the everyday memory aid, such as a string tied around the finger as a reminder, a shopping list, a calendar, or a computer (Hutchins 1999, 126, 127; Norman 1991, 17–20). *Artifacts* act as *mediators* between us and the world (Norman 1991, 22). The person-artifact interaction can be regarded as a form of *distributed cognition*, which is not just a division of labor; in fact, the interaction itself creates the possibility of *cognitive change*, creating something *qualitatively* different (O'Malley and Draper 1992, 88).

To get a better grasp of the link between Forms and models, it is necessary to take a step back from the models themselves. The technology of modeling (Odenbaugh 2008, 516) produces a layer to facilitate the understanding of a particular phenomenon. We need to remove temporarily this layer in order to be able to see more clearly, by way of what is different after removal, those features of it that can help us to connect scientific models with the Forms. When a phenomenon is analyzed with no use of models, the investigation can be developed via abstract direct representation, ADR. This kind of representation, as its name suggests, renders *directly* the object of study (Weisberg 2007, 215). An example of ADR is given by Weisberg, who cites Mendeleev's Periodic Table, which is the result of the application of ADR on each of the elements which compose it. Weisberg stresses that Mendeleev's research required a process of abstraction, which was applied *directly* on the chemical phenomena. They were studied without the aid of modeling and this investigation resulted in the identification of patterns in the elements:

> Mendeleev examined elemental properties, worked out which properties were essential and which one could be abstracted away, and then constructed a representational system that elucidated important patterns and structure among the elements. This scientific activity constitutes theory construction, but not modelling. Mendeleev represented chemical phenomena *directly*, without the mediation of a model. (Weisberg 2007, 215)

This very brief examination of ADR helps us to see its main feature: ADR allows the investigation of a particular phenomenon in an *immediate* way. By contrast, scientific models are a *medium*, through which phenomena are studied *indirectly*.

This sketch of how ADR works, eases our comprehension of what models are, via the contrast with ADR. In fact, now we know that models facilitate the examination of a phenomenon in an indirect, mediate, way. This characteristic is very important to discover whether they can be associated with Plato's Forms and to begin to understand the nature of this association. I am not saying that what we have just seen about ADR can be considered an exhaustive account of the difference between this method of investigation of a phenomenon and scientific modelling. This kind of account would not be helpful to clarify the analogy between scientific modeling and Plato's Forms. In this context, I am not going to take a position with respect to the exegetical realm centered on the interpretation of Plato's Forms. I want to focus my attention on those main characteristics of the Forms, that can be extended to every Platonic

Form, because they will gradually help us to grasp the association between scientific models and Forms and this association is crucial to understand the *epistemic function* of the Forms. To do this, I will keep as referent the excerpt of the *Symposium* quoted below, which presents the characteristics of the Form of Beauty, that are common to *all* the Platonic Forms. On these traits the analogy between Forms and scientific models will be constructed.

> This is what it is to go aright, or be led by another, into the mystery of Love: one goes always upwards for the sake of this Beauty, starting out from beautiful *things* and using them like *rising stairs*: from one body to two and from two to all beautiful bodies, then from beautiful bodies to beautiful customs, and from customs to learning beautiful things, and from these lessons he arrives in the end at this lesson, which is learning of this very Beauty, so that in the end he comes to know just *what it is to be beautiful*. (*Symposium*, 211c–d. My emphasis)

In this excerpt Plato explains clearly that you do not abstract the Forms *directly* from the empirical. Similarly, you cannot reach them *directly* in an intelligible way but you need the empirical stairs to be introduced to the realm of the non-tangible where the Forms are. The distinction between ADR and modeling that we have outlined helps us now to begin to draw the analogy between the Forms and scientific modeling. The Forms do not function as an ADR of the empirical or the purely intelligible. The Forms are not like the properties of the elements studied by Mendeleev: they cannot be abstracted directly. The Forms are *abstracta* but they are *not direct* abstractions. The Forms are the *media* used by Plato to investigate the purely intelligible, where the tangible finds its real significance, what really is. Thus, the Forms have to be associated with the models since, as we have seen, the models are the layer constructed to facilitate the investigation of a phenomenon in an indirect way. The Forms are the technology, the *cognitive artifact*, elaborated by Plato to ease the readers into the inquiry of the purely intelligible.

In Chap. 2, Sect. 2.3. I stressed that theoretical childhood is the phase of our cognitive development in which we still need the *mediation* of the written reminder to progress intellectually. Now, I have just stated that the Forms act, like models, as a *medium* to investigate the intelligible. This does not mean, however, that a connection between the Forms and modeling, and the consequent link with a mediated form of knowledge, leads the Forms to be related exclusively to theoretical childhood, an epistemic moment in which you are not yet fully aware of their function. The Forms are also connected to a more advanced kind of knowledge, because their mediation is instrumental to the investigation of the purely intelligible, which is the cognitive goal of theoretical adults.

5.1.1 The Forms and Mental Models

We have started to articulate the connection between Plato's Forms and scientific models, using the principal traits that distinguish modeling from ADR. To get a better sense of the analogy between Plato's Forms and modeling, I am going to take into consideration *mental models*. There is no agreed definition of what is a mental model (Rogers and Rutherford 1992, 289) but, for my present purposes, it is not necessary to investigate the different conceptions of mental models. I will refer only to the notion of mental models elaborated by Johnson-Laird (1983, 1988).[1]

Mental models, for Johnson-Laird, make us understand the mental processes that occur in *deductive inference* (Johnson-Laird 1983, 23–24): mental models show us what we can use to reason deductively (Johnson-Laird 1988, 226–227). In a deduction, the conclusion does *not* contain more semantic information than the premises, the conclusion of a deduction does *not* rule out some additional state of affairs over and above those ruled out by the premises: in a valid deduction, its conclusion is true in *any* situation in which the premises are true (Johnson-Laird 1988, 219). An example of a *deduction* is given by Johnson-Laird, who asks his readers to imagine the following scenario (Johnson-Laird 1983, 23. My emphasis):

> Person A asks: Where's the university?
> Person B replies: Some of those people are from there.
> Person A goes up to the group of people indicated by B and asks them the same question.

A's behavior depends on a chain of inferences that includes at its center the following *deduction*:

> Some of those people are from the university. Any person from the university is likely to know where the university is.
> ∴ Some of those people are likely to know where the university is.

Johnson-Laird takes into consideration one way in which a valid inference, like the one that we have just seen, can be made: you can imagine the situation described by the premises, then you formulate an informative conclusion which is true in that situation, and finally you consider if there is any way in which the conclusion could be false (Johnson-Laird 1988, 227). To imagine a situation is, according to Johnson-Laird, to construct a *mental model* along the lines suggested by Kenneth Craik:

> ... If the organism carries a '*small scale model*' of *external* reality and of its own possible actions *within* its head, it is able to try out various alternatives, conclude which is the best of them, react to future situations before they arise, utilize the knowledge of past events in dealing with the present and future, and in every way to react in a much fuller, safer, and more competent manner to the emergencies which face it. (Johnson-Laird 1983, 3, 1988, 215; My emphasis)

We are going to try to understand how a *mental model* can be constructed, following the different steps of the process of *deduction*, as they are described by Johnson-Laird. First of all, he considers a method for *externalizing* the *process of deduction* (Johnson-Laird 1983, 94–97). He asks his readers to suppose to want to draw a conclusion from the premises:

> All the artists are beekeepers
> All the beekeepers are chemists

One way in which to proceed is to employ a group of actors to construct a "tableau" in which some of them act as artists, some as beekeepers, and some as chemists. To represent the first premise, every person acting as an artist is also instructed to play the part of a beekeeper, and, since *the first premise is consistent with there being beekeepers who are not artists*, that role is assigned to other actors, who are told that it is uncertain whether or not they exist. In short, a tableau of the following sort is set up (Johnson-Laird 1983, 94–95):

    artist = beekeeper
    artist = beekeeper
    artist = beekeeper
        (beekeeper)
        (beekeeper)

There are three actors playing the joint roles, and two actors taking the part of *the beekeepers who are not artists*: the *parentheses* designate a directorial device establishing that the latter *may or may not exist*. Obviously, the number of actors playing the different roles is entirely arbitrary. The tableau is easily extended to accommodate the second premise (Johnson-Laird 1983, 95):

All the beekeepers are chemists.

Those actors playing beekeepers are instructed to take on the role of chemists, and an arbitrary number of new actors are introduced to play the role of *chemists who are not beekeepers; a type which, once again, may or may not exist* (Johnson-Laird 1983, 95):

    artist = beekeeper = chemist
    artist = beekeeper = chemist
    artist = beekeeper = chemist
        (beekeeper) = (chemist)
        (beekeeper) = (chemist)
                (chemist)

At this point, if you were asked whether it followed that *All the artists are chemists*, you could readily inspect the tableau and determine that this conclusion is indeed true. All that you have done is to *externalize* and to combine the information in two premises. But *deductive inference requires more than just the construction of an integrated representation of the premises: it calls for a search for counter-examples* (Johnson-Laird 1983, 95).

Johnson-Laird provides a more complicated example to illustrate this point (Johnson-Laird 1983, 95–97); he takes into consideration the case in which a troupe of actors has to represent the premises:

> None of the authors are burglars.
> Some of the chefs are burglars.

The first premise is represented by *two distinct groups* acting as the *authors* and the *burglars*, and they are instructed that they are *never* allowed to take on each other's role. The two groups are *fenced off from each other* (Johnson-Laird 1983, 95–96):

> author
> author
> author
> ---------------------------------------
> burglar
> burglar
> burglar

The tableau indicates that *no author is identical to any burglar*. The information in the second premise, *Some of the chefs are burglars*, is added by extending the tableau in a straightforward way, where *some* is taken to mean *at least some* (Johnson-Laird 1983, 96):

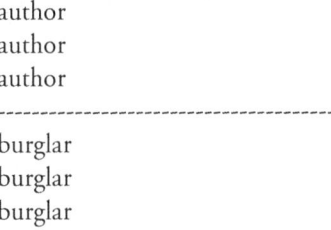

It is tempting, at this point, to conclude that *None of the authors are chefs*, or conversely, that *None of the chefs are authors*. Both conclusions are certainly consistent with the tableau. But neither is warranted, because there is another way of representing the premises (Johnson-Laird 1983, 96):

```
author
author
author            =            chef
--------------------------------------------------
          burglar   =   chef
          burglar   =   chef
          (burglar)
```

This interpretation is wholly consistent with the premises (*None of the authors are burglars*, and *Some of the chefs are burglars*), yet it invalidates the previous conclusions and suggests instead that *Some of the authors are not chefs*. However, there is still another possibility. There could be several chefs who are not burglars, and each author could be identical to such a chef (Johnson-Laird 1983, 96–97):

```
author      =            chef
author      =            chef
author      =            chef
--------------------------------------------------
          burglar   =   chef
          burglar   =   chef
          (burglar)
```

The tableau is still consistent with the premises, what it shows is that the conclusion that *Some of the authors are not chefs* is invalid. There is no other assignment of roles that is compatible with the premises, and it could therefore be supposed that there is no valid conclusion. But *in all three of the tableaux it is the case that at least Some of the chefs are not authors. This conclusion is accordingly valid* (Johnson-Laird 1983, 97).

With the idea of employing actors to take on different parts, it is obviously suggested only a hypothesis about how people might actually *make inferences*. An alternative, proposed by Johnson-Laird, to the arrangement of an external tableau, is the construction of a *mental model* (Johnson-Laird 1983, 97). A *mental model* is an *internal tableau* containing *elements that stand for the members of sets* in the same way that the actors did. Johnson-Laird indicates a general procedure for *making inferences* in this way; this general procedure is based on three main steps (Johnson-Laird 1983, 97–101).

The *first step consists in the construction of a mental model of the first premise.* The representation of a universal affirmative assertion has the following structure (Johnson-Laird 1983, 97):

All of the X are Y: x = y
      x = y
       (y)
       (y)

where *the number of tokens corresponding to x's and y's is arbitrary,* and the *items in parentheses* represent *the possible existence of y's that are not x's.* The representations of other forms of premise are straightforward (Johnson-Laird 1983, 97–98):

Some of the X are Y: x = y
       x = y
      (x) (y)

None of the X are Y: x
       x
     ----
        y
        y

Some of the X are not Y: x
        x

       (x) = y
        y

It is important to notice that *each premise requires only a single mental model.* In fact, *a crucial point about mental models* is that the system for constructing and interpreting them must embody the knowledge that the *number of entities depicted is irrelevant to any inference that is drawn* (Johnson-Laird 1983, 98).

In the *second step* of the general procedure for making inferences, *the information in the second premise is added to the mental model of the first premise, taking into account the different ways in which this can be done* (Johnson-Laird 1983, 98). The principle that motivates the search for alternative ways of adding the information from the second premise is that in a *valid deduction*, the *inference is valid* if and only if there is *no* way of interpreting the premises that is consistent with a *denial* of the conclusion (Johnson-Laird 1983, 98).

For some inferences, there is only *one* possible integrated model. For example, a premise of the form *Some of the A are B* yields the model (Johnson-Laird 1983, 98):

$$a = b$$
$$a = b$$
$$(a)(b)$$

and a second premise of the form *All of the B are C* can be *integrated only* by forming the model:

$$a = b = c$$
$$a = b = c$$
$$(a)(b) = c$$
$$(c)$$

from which it follows that *Some of the A are C*. There is *no alternative model of the premises that violates this conclusion* (Johnson-Laird 1983, 99).

For other inferences, it is necessary to construct and evaluate two models. For example, premises of the form (Johnson-Laird 1983, 99):

Some of the A are not B
All the C are B

can be represented by the model:

```
a
a
-------------------
(a) = b = c
      b = c
      (b)
```

In this model, the conclusion *Some of the A are not C* is *true because the A's above the broken line are not C's*, and the *converse conclusion Some of the C are not A is also true* because *there is at least one C that is not linked to an A.* The search for an alternative model to falsify these putative conclusions yields a second model (Johnson-Laird 1983, 99):

```
a
a
--------------------
  (a) = b = c
  (a) = b = c
      (b)
```

This model *rules out the second conclusion because all the C's are A's*, but there is *no way to destroy the first conclusion, Some of the A are not C, which is accordingly valid.*

Still *other premises yield three different models* (Johnson-Laird 1983, 99). For example, the premises:

All of the B are A
None of the B are C

yield the model:

```
c
c
-------------------
      b = a
      b = a
        (a)
```

which suggests the conclusion *None of the C are A*, or its converse *None of the A are C*. The search for a counter-example yields the model (Johnson-Laird 1983, 100):

```
c
c = a
-------------
   b = a
   b = a
      (a)
```

which falsifies both of these conclusions and suggests instead that *Some of the C are not A* or conversely that *Some of the A are not C*. The search for a counter-example to these conclusions yields:

```
c   =   a
c   =   a
--------------------
   b = a
   b = a
      (a)
```

which shows that *the only conclusion Some of the A are not C is valid*. The last model by itself, of course, suggests the invalid conclusion that *All of the C are A*, but that conclusion is ruled out by the previous models. Although these premises have three models, an entirely feasible strategy, illustrated here, is to construct one model and then to try out various modifications of it that are consistent with the premises (Johnson-Laird 1983, 100).

In the *third and last step of the procedure for making inferences, a conclusion has to be framed to express the relation, if any, between the 'end' terms that holds in all the models of the premises*. An 'end' term is one which occurs in only a single premise, unlike the 'middle' term which occurs in both premises. "*If there is no such relation between the end terms*, the only valid conclusions that can be drawn are trivial ones, such as conjunction or disjunction of the premises, and subjects generally respond that there is *no valid conclusion*" (Johnson-Laird 1983, 101. My emphasis).

As we have just seen, a fundamental step in the construction of mental models is the search for an alternative model of the premises that falsifies the conclusion. If there is no such model, then the conclusion is valid; if you cannot find such a model but your search is not exhaustive, the conclusion *may* be valid; if you find such a model, then your conclusion is not valid, and you must consider the new model together with any previous ones to see whether they support a new conclusion, and then in turn test that conclusion, and so on (Johnson-Laird 1988, 228). Obviously, the greater the number of different models that have to be constructed to draw a valid inference, the harder the task will be, as exemplified by this difficult inference, examined by Johnson-Laird (1988, 228–231):

> None of the archaeologists is a biologist.
> All the biologists are chess-players.

Given that these assertions are true, what, if anything, can follow validly? The right answer for the right reasons calls for the construction of at least three models (Johnson-Laird 1988, 229). This is the first model:

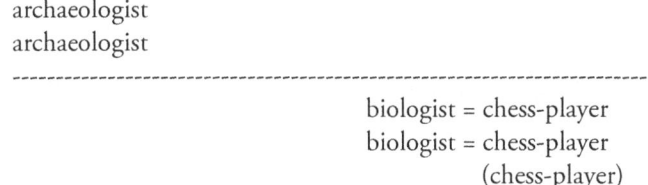

The reasoner imagines some arbitrary number of archaeologists, and demarcates them as not identical to biologists (as indicated here by the barrier). The second premise calls for each biologist to be identified as a chess-player. Of course *there may be chess-players who are not biologists.* They are represented by the *token within parentheses*, and it is assumed that they are initially put on the same side of the barrier as the biologists: *people do not realize at once that such chess-player could be an archaeologist* (Johnson-Laird 1988, 229). This first model yields an informative conclusion:

None of the archaeologists is a chess-player

If the model is scanned in the opposite direction to which it was constructed, a procedure that is relatively difficult, it yields the converse conclusion:

None of the chess-players is an archaeologist

Neither of these conclusions is valid, since they can be refuted by a second model (Johnson-Laird 1988, 229):

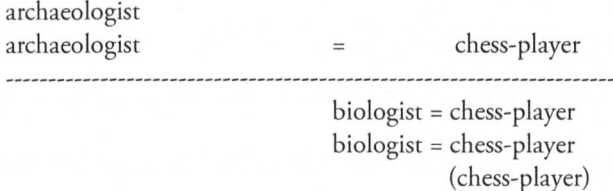

The two models together support the informative conclusions (Johnson-Laird 1988, 230):

Some of the archaeologists are not chess-players
Some of the chess-players are not archaeologists

Finally, a third model refutes the first of these conclusions:

archaeologist   =   chess-player
archaeologist   =   chess-player
----
              biologist = chess-player
              biologist = chess-player
                      (chess-player)

Granted that people tend to scan models in the direction in which they construct them, *reasoners who have got this far have moved from a model in which none of the archaeologists is a chess-player to one in which all the archaeologists are chess-players* (Johnson-Laird 1988, 230). They may well respond

> There is no valid conclusion

Indeed, *if the models are scanned in the opposite direction, there is a conclusion that holds in all three*, which is *the only valid conclusion* relating the two terms (Johnson-Laird 1988, 230):

> Some of the chess-players are not archaeologists

As Johnson-Laird points out, "*vision* yields mental models" (Johnson-Laird 1988, 231. My emphasis). We have noted (Chap. 3, Sects. 3.1 and 3.2), thanks to Giaquinto's research, the important *epistemic role* of *visualization*, which is not a mere source of evidence, but it is what triggers the belief-forming dispositions, that render possible new discoveries. The vision involved in the formation of mental models has clearly *not* an evidence-providing role: we do not construct mental models because we want to know how authors or archaeologists look like, we construct mental models to ease a process of *deductive inference*. Mental models point to an *epistemic* role of vision.

The *epistemic* importance of mental models is reinforced, if we consider them in the context of *problem solving*. Larkin and Simon (Chap. 4, Sect. 4.1) helped us to reflect on the fact that both external and internal diagrammatic representations play the same role in *problem solving*: "*mental images* play a role in *problem solving* quite analogous to the role played by external diagrams" (Larkin and Simon 1995, 105. My emphasis): mental images, as well as external diagrams, facilitate problem-solving *inferences*. Johnson-Laird has just showed us a method for the *internalization* of the process of deduction, the construction of mental models. This method is much more convenient than the *externalization* of the process of deduction that consists in the construction of a tableau in which actors play different parts, corresponding to the different subjects involved in the inference. Mental models ease the deductive process, contributing in this way to problem solving.

We have seen (Chap. 4, Sect. 4.2.1) that *reasoning problems* have been posed via *Hyperproof*. It has been suggested that the schematization used by Plato in the *Republic* to represent the different steps of cognitive progress (*Republic*, VI 509d–511) has the same *function* of Hyperproof: both propose *logic* problems that the readers have to solve in order to improve their *general reasoning abilities*. Mental models increase our *logical competence*, improving our capacity for making inferences. The concept of vision related to mental models has an *epistemic* rather than an *empiric* character. We know that, for Plato, the truth can be found only in the realm of the *intelligible*. Thus, mental models are the *media* that, in a Platonic perspective, lead us from an inferior *cognitive* state, that I have labeled as theoretical childhood, to an advanced epistemic phase, that I have called theoretical adulthood (Chap. 2, Sect. 2.3). Mental models can be considered *analogues of* the Platonic Forms because of their *epistemic function*: they are both *media* in which *vision* has an *epistemic* role, increasing our cognitive ability in *problem solving*. This emphasis on the kind of *vision considered epistemically* is provided by Plato himself, who chooses the term Idea, also rendered in translation as Form, to point to what is pertinent to the realm of the intelligible. Ideas are intelligible (τὰς δ’ αὖ ἰδέας νοεῖσθαι μέυ *tas d au ideas noeisthai men Rep*, VI 507b): the word *Idea* comes from the root *-id*, that is found in the verbal form *eidon*, aorist of the verb *oraō*. The meaning of this verb comprises of a metaphysical shade: *oraō* means *mental* sight (Liddell et al. 1996, 1245). Thus, Plato chooses to refer to the Forms using a word which is *etymologically related to a concept of vision cognitively tainted*. The vision of the Ideas is a vision that makes you *know*, connecting you with the intelligible realm. Having in mind the relation between *theōreō and oraō*, and the connection between *cognition and vision*, I have chosen, as we have seen (Sect. 2.3), to label the two phases of my reconstruction of cognitive development in Plato, theoretical childhood and theoretical adulthood.

Plato, describing the stages of human cognitive evolution (*Republic*, VI 509d–511), poses a *logic* problem to stimulate the development of the general reasoning abilities of his readers. A crucial step of this cognitive progress is the moment in which Plato's readers pass from an inferior phase of rational growth, that I have called theoretical childhood, to a superior phase of cognitive development, that I have labeled as theoretical adulthood. This fundamental passage is rendered possible, in my interpretation of Plato, by the Forms. I associate the Forms with mental models because of their *function*. Both the Forms and mental models are not abstract direct representations: they are *not* abstracted *directly* from the *empirical or the intelligible realm*. They are the *cognitive artifacts*, the mediators (Sect. 5.1), that lead us toward the intelligible, pointing to an *epistemic role of vision*, as means of cognitive progress. If we recognize this *epistemic* role of vision, and the related Platonic exhortation to rational development, we realize that Plato's words are just a *means* devised to make us exercise our reasoning abilities. This realization makes us become the kind of spectacle lovers who are able to grasp the "truth beyond the shadow play," as Mattéi has pointed to us (Chap. 2, Sect. 2.1, Mattéi 1988, 79).

Notes

1. The examination of the criticisms of this notion of mental models (see, e.g., Stenning 2002, 115) is beyond the scope of the present research.

References

Texts and Translations

Plato. Republic. 1997. Translated by Grube, G.M.A. Revised by Reeve, C.D.C. In *Plato: Complete Works*, ed. J.M. Cooper. Indianapolis: Hackett.
Plato. Symposium. 1997. Translated by Nehamas, Alexander and Woodruff, Paul. In *Plato: Complete Works*, ed. J.M. Cooper. Indianapolis: Hackett.

Recent Works

Hutchins, Edwin. 1999. Cognitive Artifacts. In *The MIT Encyclopedia of the Cognitive Sciences*, ed. Robert A. Wilson and Frank C. Keil. Cambridge, MA: MIT Press.
Johnson-Laird, Philip N. 1983. *Mental Models: Towards a Cognitive Science of Language, Inference, and Consciousness*. Cambridge: Cambridge University Press.

———. 1988. *The Computer and the Mind: An Introduction to Cognitive Science.* Cambridge, MA: Harvard University Press.

Larkin, Jill H., and Herbert A. Simon. 1995. Why a Diagram Is (Sometimes) Worth Ten Thousand Words. In *Diagrammatic Reasoning: Cognitive and Computational Perspectives*, ed. Janice Glasgow, N. Hari Narayanan, and B. Chandrasekaran. Cambridge, MA: MIT Press.

Liddell, Henry G., Robert Scott, Henry Stuart Jones, and Roderick McKenzie. 1996. *A Greek-English Lexicon.* Oxford: Clarendon Press.

Mattéi, Jean-François. 1988. The Theatre of Myth in Plato. In *Platonic Writings/ Platonic Readings*, ed. Charles L. Griswold. New York: Routledge.

Norman, Donald A. 1991. Cognitive Artifacts. In *Designing Interaction: Psychology at the Human-Computer Interface*, ed. John M. Carroll. Cambridge: Cambridge University Press.

O'Malley, Claire, and Steve Draper. 1992. Representation and Interaction: Are Mental Models all in the Mind? In *Models in the Mind: Theory, Perspective and Application*, ed. Yvonne Rogers, Andrew Rutherford, and Peter A. Bibby. London: Academic Press.

Odenbaugh, Jay. 2008. Models. In *A Companion to the Philosophy of Biology*, ed. Sahotra Sarkar and Anya Plutynski. Oxford: Blackwell Publishing.

Rogers, Yvonne, and Andrew Rutherford. 1992. Future Directions in Mental Models Research. In *Models in the Mind: Theory, Perspective and Application*, ed. Yvonne Rogers, Andrew Rutherford, and Peter A. Bibby. London: Academic Press.

Stenning, Keith. 2002. *Seeing Reason: Image and Language in Learning to Think.* Oxford: Oxford University Press.

Weisberg, Michael. 2007. Who is a Modeler? *British Journal for Philosophy of Science* 58 (2): 207–233.

6

Theoretical Adulthood and Structuralism

Abstract Shapiro's *ante rem structuralism* is a kind of structuralism that ignores the *individual properties of the objects*, that are irrelevant, and it considers only the objects as *positions* in a structure. The axioms governing these objects do *not* assert *definite* truths but they *define* a kind of structure of mathematical interest. The axiomatic approach connected to structuralism can be thus related to the axiomatic approach that has been called as bottom-up, based on premises which can be questioned in light of the results obtained. This axiomatic approach has been associated with the level of mathematical complexity pertinent to the investigations of theoretical adults.

Keywords Mental number lines • *Ante rem* structuralism • Realism in ontology • Naturalized epistemology

6.1 Number Lines and Structures

We have considered what could be the purpose of the Platonic choice of *a* schematization to render the stages of human cognitive development. In this last part of my research, I want to work on what could be the

S. Saracco, *Plato, Diagrammatic Reasoning and Mental Models*,
https://doi.org/10.1007/978-3-031-27658-3_6

77

reasons why Plato has chosen *that* diagram, a line segment subdivided into sectors, to stimulate us intellectually. As we have seen (Chap. 4, Sect. 4.1.1), diagrams are useful aids to problem solving: it is more common to teach *arithmetical operations* in terms of their *spatial analogues* than in terms of direct logical definitions. It is often found useful to think of numbers as forming a spatially ordered series, along which something can move in either direction. So, for a child who is learning to understand numbers, it is helpful to think of subtraction as moving backward a fixed number of steps (Sloman 1995, 17). Also Giaquinto points to the importance of *visual number lines* in our *mathematical thinking* (Giaquinto 2007, 111). The *visual argument* is persuasive and makes the correctness of the proposition obvious in a *direct* way (Giaquinto 2007, 114): whole number addition can be easily represented by "a movement to the right from the position marking one addend by the length representing the other addend, the result being represented by the end position (or the length of the segment from the origin to the position). Whole number subtraction $n-k$ can be represented as a leftward movement from the position representing n by the length representing k, the result being represented by the end position….We also have representations of multiplication, division, and rational numbers in terms of the number line" (Giaquinto 2007, 111). The *epistemic* result is achieved by deploying one's implicit grasp of these facts of representation together with vision or visual imagination and some simple deduction (Giaquinto 2007, 115).

This association of *numerals* with a *spatial line* can make us think that Plato could have chosen to represent the human cognitive growth with a line, to suggest an involvement of mathematics in this process of epistemic development. Assuming that this could be true, new questions arise: *where* should mathematics exercise its power? Within the four subsegments of cognitive progress mentioned by Plato or could there be a more advanced phase of intellectual development where mathematics can express its full potential?

A help for the progress of this investigation is given by Giaquinto's work. He takes into consideration the *mental number line* that allows us by means of visual representation to know an *infinite structure*, the structure of the natural numbers. Giaquinto, working on *visual cognition of an infinite structure*, refers to the "structure of the finite cardinals under their natural 'less than' ordering. This structure, which I will call 'N', is shared by the set of arabic numerals of the decimal place system in their standard ordering" (Giaquinto 2007, 226; See also Giaquinto 2008, 53).

As Giaquinto notices, "an obvious problem with the idea that a *mental number line* provides a grasp of the *natural number structure* is that we cannot see or visualize more than a finite part of any such line" (Giaquinto 2008, 53. My emphasis).[1] Thus, according to Giaquinto, when it comes to actual images (or percepts) something like the figure below will be the best we can do (Giaquinto 2008, 53) (Fig. 6.1).

The fact that we cannot see or visualize more than a finite fragment of any instance of an infinite structure is not an insurmountable obstacle. For Giaquinto there are two kinds of visual representations, visual category specification and visual image. "A *visual category specification* is a set of related feature descriptions stored more or less permanently; a *visual image* is a fleeting pattern of activity in a specialized visual buffer, produced by activation of a stored category specification. What is impossible is an infinitely extended visual image. But it is possible, and not at all puzzling, that *a category specification specifies a line with no right end, one that continues rightward endlessly*" (Giaquinto 2007, 227. See also Giaquinto 2008, 54).

In having a visual category specification for the mental number line, "we have a grasp of a *type* of *structured set*, namely a set of number marks on a line endless to the right taken in their left-to-right order of precedence. Secondly, we can have knowledge of the *structure* **N** as the structure of a 'number line' of this type" (Giaquinto 2008, 56. My emphasis. See also Giaquinto 2007, 228). Giaquinto has pointed to the importance of mental number lines for the cognition of infinite *structures* such as the natural number *structure*. We are going to see what *structuralism* is and what could be its relation to Plato's philosophy.

Fig. 6.1 The infinity of the natural number structure can be rendered via a mental number line with no right end

6.1.1   Structuralism and Plato[2]

Giaquinto has just showed us how the infinity of the natural number structure can be rendered via a mental number line with no right end. This representation abstracts away from the nature of the objects, the natural numbers, which instantiate the natural number *structure*. According to *structuralism, numbers*, e.g., in the natural number structure, should be treated as *positions in structures*. For the structuralist, "mathematics is seen as the investigation…of 'abstract structures', systems of objects fulfilling certain structural relations among themselves and in relation to other systems, without regard to the particular nature of the objects themselves.…the 'objects' involved serve only to mark 'positions' in a relational system; and the 'axioms' governing these objects are thought of, *not* as *asserting definite truths*, but as *defining* a type of structure of mathematical interest" (Hellman 2005, 536–537). We will come back to Hellman's words shortly. Now, I want to take into consideration a particular instance of structuralism, Shapiro's *ante rem structuralism*. The basics of this kind of structuralism are well explained by Sereni:

> Arithmetic assertions…are not centred on particular objects…Rather, they are based upon the *positions* of the progression structure. For example, the assertion '3 < 5' does not state that a particular object, 3, is in the relation 'being minor of' with another particular object, 5. Rather, it states that the *position* of the progression structure that we call '3' (that will be the third or fourth *position* of the structure, according to the fact that we choose to make the structure begin with 1 or 0) comes before, according to the order relation that exists among these *positions*, the *position* of that same structure that we call '5'. *The fact that exist particular objects, numbers, or other abstract objects, or concrete objects, that occupy those positions and that constitute a system that exemplifies the structure in question, is something that lies outside the object of arithmetic and the significance of its assertions. There could exist natural numbers, occupying the positions that we call with their names;…or there could exist nothing that satisfies the relations of the progression structure. Independently from this, the object of arithmetic-that specific structure- does not change, and its theorems remain true descriptions of that object.* (Sereni 2020, 166–167. My translation. My emphasis)

These words have helped us to understand what *ante rem* structuralism is: it is a kind of structuralism that *ignores* the *individual properties of the objects*, that are *irrelevant*, and it considers only an object as a *position* in a structure. At the base of *ante rem* structuralism there is *abstraction* (Shapiro 1997, 74). Plato, in the *Republic*, specifies that mathematics has not to be used as retailers and tradesmen do, just to be able to buy and sell, but it must be used to discuss the *nature of the numbers*, moving in this way the soul from *becoming* to truth and *being* (*Republic*, VII 525b–c.). Thus, for Plato, *abstraction* is the core of a correct use of mathematics: we have to *abstract* away from the empirical to turn our rational attention to the intelligible realm, to which the *nature* of the numbers is pertinent. This importance of abstraction, and the consequent interest in the universal rather than the empirical, is what connects *ante rem structuralism* with Plato's philosophy, as it is confirmed by Shapiro himself who, in his *Thinking about Mathematics: The Philosophy of Mathematics* (Shapiro 2000, 58–60), considers an excerpt taken from the *Philebus* (56d–e. My emphasis):

> Don't we have to agree, first, that the *arithmetic of the many* is one thing, and *the philosophers' arithmetic* is quite another?…*First* there are *those who compute sums of quite unequal units*, such as two armies or two herds of cattle, regardless whether they are tiny or huge. But then there are *the others* who would not follow their example, unless it were *guaranteed that none of those infinitely many units differed in the least from any of the others.*

In this passage Plato emphasizes the difference between ordinary arithmetic and philosopher's arithmetic. As Shapiro notices, the "philosopher's arithmetic applies precisely and strictly only to the world of Being" (Shapiro 2000, 58). Numbers are studied in different ways by philosophers and non-philosophers: "the philosopher's numbers are numbers of pure units" (Shapiro 2000, 59). When the philosophers count, as the lines above explain to us, they take into consideration the *essence* of the units involved in the process of counting. The calculation of the philosophers takes place within the realm of Being, mentioned by Shapiro, where there is no difference among the units of the calculation: "Plato's arithmetic is a part of higher philosophy, where one comes to grasp the metaphysical nature of number itself" (Shapiro 2000, 60). When the philosophers count, the units involved in this process are the same because the philosopher examines the metaphysical facet of them, their essential aspect and not their contingent appearance. The philosophers count what is essentially homogenous. Differently from the ordinary arithmetician, they know that the heterogeneity of the sensible side of the units counted has to be overcome.

Shapiro points to the fact that "*ante rem* structuralism is a variant of traditional Platonism" (Shapiro 2011, 130. See also Shapiro 2006, 142): *ante rem* structuralism is an instance of the view that he calls 'realism-in-ontology' (Shapiro 2006, 142). In Shapiro's structuralism there is an "*existential* commitment to both *structural universals* and their *positions*. The structural universals so described are '*ante rem*' because, *like Plato's Forms, they exist independently of the systems that exemplify them*" (MacBride 2008, 156. My emphasis). The "*ante rem* structuralist takes a Platonic view of structures: they exist and are available for mathematical description as complex objects in their own right, *whether or not exemplified by any independent collection of objects*" (Wright 2000, 330. My emphasis).

Plato's Forms have been associated with mental models (Chap. 5, Sect. 5.1.1). This analogy has been based on the *epistemic function* of Plato's Ideas and mental models: they are the *media* that improve our *cognitive vision*. They are *not* abstract direct representations of the empirical or the intelligible; they are the *mediators*, the *cognitive artifacts* that turn our rational attention to the realm of the intelligible. This does *not* mean that the Forms do *not* exist independently of the objects that exemplify them. Plato's Forms can make us see with *epistemic* eyes what it is to be beautiful, even if there are not anymore empirical objects that are beautiful. In the same way, mental models would help us to make correct inferences even if the subjects of those inferences, such as artists or beekeepers, would not exist. The *cognitive function* of Plato's Forms and mental models is *independent* of the existence of the subjects to which they refer. Even so, they are not abstract *direct* representations of the intelligible because if they were *exclusively* pertinent to the realm of the intelligible, we would not be able to grasp them. Mental models that refer neither to identifiable subjects, such as artists, nor, at the least, to a, b, or c are not conceivable. A Form that is not a Form *of*- a Form *of* beauty, or of something else, is not conceivable. Nevertheless, this does not mean that the Form as *cognitive artifact* would not exist if there were *no* beautiful things. Thus, Plato's Forms and *ante rem* structuralism can be associated.

Shapiro connects *ante rem* structuralism with Plato's philosophy: for Plato truth is disentangled from the *empirical* realm and can be found in the *purely intelligible*; in the same way, for Shapiro, it is irrelevant the *empirical* existence of objects that exemplify the structures that he is taking into consideration; these objects exist *ontologically*, as those *positions* in a structure that can be grasped via an act of *intellection. Both for Shapiro and for Plato, the truth is not in the empirical but in the intelligible realm.* The existence of the structures is posited by Shapiro via an axiomatic theory of structures. Shapiro's structures are axiomatically characterized (Sereni 2019, 253); nevertheless, Hellman, as we have seen, has clarified that the axioms, governing the objects that in structuralism are positions in a structure, do *not* assert *definite* truths but they *define* a kind of structure of mathematical interest (Hellman 2005, 537). The axiomatic approach connected to structuralism can be thus related to the axiomatic approach that in Chap. 2, Sect. 2.3 has been called as bottom-up: there are not axioms, that are never questioned, used to logically derive mathematical truths from them; on the contrary, there are axioms whose truth can be reconsidered in light of the results of the mathematical problem examined. This is an axiomatic approach proper of a higher-level of mathematical complexity, pertinent to the investigations of theoretical adults who, as we have seen, analyze the purely intelligible. Recall, we have distinguished between two levels of mathematical complexity, the first level, "the method of geometry and mathematics in general" (Heath 1921, 290), was associated with an axiomatic approach that we defined as top-down axiomatic approach: with this method, results are logically deduced from unquestioned axioms. This level of mathematical complexity is useful to turn our rational attention from the tangible to the intelligible. This focus on the intelligible is for Plato fundamental to evolve intellectually till to the point in which we become theoretical adults. The mathematics utilized by theoretical adults is based on a bottom-up axiomatic approach. At this level of sophistication, the consequences of the problem have to be utilized to reconsider the truth of the premises.

The distinction of two levels of mathematical complexity, one related to the top-down axiomatic approach, proper to theoretical children, the other connected with a bottom-up axiomatic approach, proper to theoretical adults, makes us realize the importance of mathematics in Plato. Mathematics, as Foley (Chap. 2, Sect. 2.3) has helped us to notice, is what leads us from concern for physical objects to eternal objects (Foley 2008, 12). A higher-level of mathematical complexity, represented by Shapiro's *ante rem* structuralism, makes us focus our cognitive eyes on the intelligible, where structural universals are. These considerations on mathematics bring us back to our reflections in Sect. 6.1: Plato has chosen to represent human cognitive growth with a line. The ordered numeral system is associated with a spatial line and it is common to teach arithmetical operations in terms of their spatial analogues (Chap. 4, Sect. 4.1.1). Thus, Plato has chosen to represent rational development using *that* diagram, a line segment subdivided into sections, to point to us the importance of mathematics in the process of rational growth. Mathematics permeates our entire epistemic development that, in my interpretation of Plato, goes from theoretical childhood to theoretical adulthood. It is pertinent to this advanced stage of rational progress, a kind of mathematics whose interest is the *ontological* existence of its objects, not their correspondent empirical instantiations.

I have associated the level of mathematical complexity proper to structuralism with theoretical adulthood.[3] It can be objected the existence of this phase of epistemic growth. I have never stated that the phase of superior rational development that I label as theoretical adulthood is the only way to respond to the cognitive stimulation of Plato's text. This would be contrary to the non-indoctrinative Platonic higher-order pedagogy that, as we have seen (Chap. 2, Sect. 2.2), presents to the reader what Plato's idea of truth is, but it does not impose the acceptance of this truth. According to my hermeneutic approach, the words of Plato's dialogues are meant to stimulate cognitively the readers. In this way, they acquire conscience of their intellectual capacities. The exercise of these cognitive skills can result in a radical criticism of Plato's idea of truth. I have accepted this idea and I have responded to the Platonic request of collaboration with his text, elaborating a new theoretical framework, characterized by two moments of epistemic growth, theoretical childhood, which corresponds to the description of cognitive development provided by Plato in the *Republic*, and theoretical adulthood, which is not the object of a direct Platonic description.

6.1.1.1 Epistemology in *Ante Rem* Structuralism: The Access Problem

As we have just seen, *ante rem* structuralism is a theory about what (mathematical) *universals* there are. Shapiro offers a stratified epistemology,[4] in which each stage corresponds to the acquisition of knowledge of successively more complex mathematical structures. Knowledge of structures begins with our capacity to recognize small, finite, instantiated patterns or structures; for example, short strings of numerals. The subject observes one or more systems of objects arranged in various ways and she *abstracts* away from the irrelevant *tokens*, apprehending the *types* (*universals*) under which they fall. This abstractionist step of Shapiro's epistemology allows the individuals to know small cardinal number structures but since our powers of perceptual discrimination are essentially limited, our ability to abstract types from tokens with which we are acquainted will not provide us with knowledge of large natural numbers structures such as the 1000 pattern. Thus Shapiro postulates the existence of a faculty of *projection*: this faculty enables us to arrange the patterns obtained by simple abstraction and *recognize* that they themselves exhibit an overarching pattern. This yields knowledge of large finite structures, and eventually knowledge of the natural number structure itself. But the faculty of projection is still too limited for mathematical purposes. To deal with still larger structures an alternative epistemological strategy is proposed: Shapiro poses the need of a formal language that provides appropriate definitions of the structures to allow us to know them. It is consequently our ability to grasp *direct descriptions* of large infinite structures that grounds our knowledge of them.

These steps of Shapiro's epistemology, according to MacBride, do not provide any answer to the problem that he defines as "the access problem" (MacBride 2008): how can mathematicians reliably access truths about an *abstract* realm to which they cannot travel and from which they receive no signals? (MacBride 2008, 155. My emphasis). For MacBride the problem consists in a tension between Shapiro's *realism in ontology* and *naturalized epistemology*: how can a *physical* being located in a physical universe know the *abstract* realm, that includes *ante rem* universals and infinite structures (MacBride 2008)? Shapiro's reply (Shapiro 2011, 149. My emphasis) to MacBride's doubts is that

> My game, again, is to provide a justification for a philosophical interpreta-
> tion of mathematics, an interpretation which includes a thesis concerning
> what mathematics is about-*ante rem* structures. This philosophical inter-
> pretation is not a deductive enterprise, where I would have to start with
> non-mathematical, self-evident premises. *This is a different game from show-
> ing a sceptic that mathematics itself is true and known.*

According to Shapiro, the goal of his research is to demonstrate that mathematical knowledge *just* is knowledge of *ante rem* structures. This has not to be proved from accepted *non*-mathematical premises. Shapiro's research aims at studying *ante rem* structures. As we have seen, these structures possess an *ontological* reality independent from the empirical existence of entities which physically instantiate them. This focus on the universal rather than the empirical realm is common to Shapiro and Plato, as Shapiro himself acknowledges (Shapiro 2006, 142; Shapiro 2011, 130). Both Shapiro and Plato do not tell us where their universal evidence comes from. But Plato has chosen to provide us with cognitive stimulations which are *entrance points* to this epistemic realm. The cognitive awareness acquired thanks to this rational stimulation gives us the chance to choose to criticize, even radically, Plato's system and every aspect that characterizes it.

Acknowledgments I would like to thank Marco Panza and Andrea Sereni for having introduced me to the subject of visual thinking and its epistemological importance. My ideas about structuralism and its connection with Plato's philosophy have been enriched thanks to Andrea Sereni's clarifications. I am indebted to Marcus Giaquinto for a very interesting dialectic exchange on visual thinking and Plato. I would like to express my very great appreciation to an anonymous referee of *Plato Journal*, whose suggestion for further development of my work on visual thinking, has resulted in the present research. The bibliographic search that has given me the chance to write this book would not have started without the generous help of Jeremy Avigad.

Notes

1. On this topic see also Giaquinto Marcus. 2007. *Visual Thinking in Mathematics: An Epistemological Study.* Oxford: Oxford University Press, chapter 11, see in particular pp. 226–236.
2. On this subject see also Saracco, Susanna. 2021. "Plato's Intellectual Development and Visual Thinking." *Plato Journal*, 21.
3. On theoretical adulthood see the fifth chapter of my book (Saracco 2017), *Theoretical Adulthood.*
4. Shapiro's epistemology is efficaciously summarized in MacBride, Fraser. 2008. "Can *Ante Rem* Structuralism Solve the Access Problem?" *The Philosophical Quarterly* 58 (230), pp. 157–158.

References

Texts and Translations

Plato. *Philebus*. 1997. Translated by Frede, Dorothea. In *Plato: Complete Works*, ed. J.M. Cooper. Indianapolis: Hackett.
Plato. *Republic*. 1997. Translated by Grube, G.M.A. Revised by Reeve, C.D.C. In *Plato: Complete Works*, ed. J.M. Cooper. Indianapolis: Hackett.

Recent Works

Foley, R. 2008. Plato's Undividable Line: Contradiction and Method in Republic VI. *Journal of the History of Philosophy* 46 (1): 1–23.
Giaquinto, Marcus. 2007. *Visual Thinking in Mathematics: An Epistemological Study*. Oxford: Oxford University Press.
———. 2008. Visualizing in Mathematics. In *The Philosophy of Mathematical Practice*, ed. Paolo Mancosu. New York: Oxford University Press.
Heath, T. 1921. *A History of Greek Mathematics*. Oxford: The Clarendon Press.
Hellman, Geoffrey. 2005. Structuralism. In *The Oxford Handbook of Philosophy of Mathematics and Logic*, ed. Stewart Shapiro. Oxford: Oxford University Press.
MacBride, Fraser. 2008. Can *Ante Rem* Structuralism Solve the Access Problem? *The Philosophical Quarterly* 58 (230): 155–164.
Sereni, Andrea. 2019. On the Philosophical Significance of Frege's Constraint. *Philosophia Mathematica* 27 (2): 244–275.

———. 2020. "Numeri, oggetti e strutture: sull'eredità contemporanea del problema dei fondamenti" [Numbers, Objects and Structures: On the Contemporary Heritage of the Foundations Problem]. In *L'Arte di Pensare: Matematica e Filosofia [The Art of Thinking: Mathematics and Philosophy]*, ed. Gabriele Lolli and Francesco S. Tortoriello. Novara: UTET.

Shapiro, Stewart. 1997. *Philosophy of Mathematics: Structure and Ontology*. Oxford: Oxford University Press.

———. 2000. *Thinking about Mathematics: The Philosophy of Mathematics*. Oxford: Oxford University Press.

———. 2006. Structure and Identity. In *Identity and Modality*, ed. Fraser MacBride. Oxford: Oxford University Press.

———. 2011. Epistemology of Mathematics: What Are the Questions? What Count as Answers? *The Philosophical Quarterly* 61 (242): 130–150.

Sloman, Aaron. 1995. Musings on the Roles of Logical and Nonlogical Representations in Intelligence. In *Diagrammatic Reasoning: Cognitive and Computational Perspectives*, ed. Janice Glasgow, N. Hari Narayanan, and B. Chandrasekaran. Cambridge, MA: MIT Press.

Wright, Crispin. 2000. Neo-Fregean Foundations for Real Analysis: Some Reflections on Frege's Constraint. *Notre Dame Journal of Formal Logic* 41 (4): 317–334.

Index[1]

---

[1] Note: Page numbers followed by 'n' refer to notes.

The manufacturer's authorised representative in the EU is Springer
Nature Customer Service Centre GmbH, Europaplatz 3, 69115 Heidelberg,
Germany. If you have any concerns regarding our products, please
contact ProductSafety@springernature.com

Printed and bound by CPI Group (UK) Ltd, Croydon, CR0 4YY
29/04/2026
02099525-0001